Design of Event-Driven SPAD-Based 3D Image Sensors

The book presents a top-down circuit description for the implementation of asynchronous vision sensors based on Single-Photon Avalanche Diodes (SPADs). It provides design considerations to convey the SPADs pulses through a channel shared by all the pixels. The book also presents architectures where dynamic vision sensors and SPADs converge.

Design of Event-Driven SPAD-Based 3D Image Sensors provides detailed technical insights about novel image sensor architectures with SPADs with asynchronous operation. At the sensor level, the book provides asynchronous circuitry to read and arbiter the pixel outputs. The authors explore new LiDAR architectures with asynchronous operation and provide insights into their design. A detailed benchmark of modern and competitive LiDAR systems is also presented. At the pixel level, the book provides design considerations to convey the SPADs pulses through a channel shared by all the pixels. At the sensor level, the book provides asynchronous circuitry to read and arbiter the pixel outputs. Finally, experimental results of very novel LiDAR systems with asynchronous operation will be provided and analyzed.

The book is written for professionals who want to explore new tendencies on the design of image sensors for the implementation of LiDAR systems.

Dr. Rubén Gómez-Merchán received an M.S. (Hons.) degree in microelectronics and the Ph.D. degree in physics-electronics from the University of Seville (US), Seville, Spain, in 2020 and 2024, respectively. From 2018 to 2019, he was with the Department of Electronic Engineering, US, where his main research interests were dc/dc converters and PV systems. In 2019, he was granted by the Spanish National Research Council (CSIC), starting his activity at the Instituto de Microelectronica de Sevilla (IMSE), US. From September 2022 to December 2022, he was a visiting researcher at Sony Europe Design Center, in Oslo, Norway. His main research interests include smart CMOS sensors and address event representation (AER) vision systems.

Dr. Juan Antonio Leñero Bardallo (Senior Member, IEEE) received an M.Sc. in telecommunications engineering and a Ph.D. in microelectronics from the University of Seville, Seville, Spain, in 2005 and 2010, respectively. After completing his Ph.D., he served in several academic institutions and worked for the semiconductor industry. He was a Postdoctoral Associate at Yale University, New Haven (USA), and the University of Oslo (Norway). Since February 2018, he has been an Associate

Professor at the University of Seville, Spain. For more than 20 years, he has been involved in designing Address Event Representation (AER) vision sensors in many application scenarios such as space instrumentation, navigation systems, surveillance, and biomedical systems.

Dr. Ángel Rodríguez-Vázquez (Life Fellow, IEEE), received a Ph.D. in physics-electronics from the University of Seville in 1982. After stays at the University of California at Berkeley and Texas A&M University, he became a Full Professor of electronics at the University of Seville in 1995. He co-founded the Instituto de Microelectrónica de Sevilla. In 2001, he was the Main Promotor of AnaFocus Ltd., and served as its CEO until June 2009, when the company reached maturity as a worldwide provider of smart CMOS imagers. He has also participated in the foundation of the Hungarian start-up AnaLogic Ltd. He has ten patents filed; AnaFocus started based on his patents on vision chip architectures. His lab designed many high-performance mixed-signal chips in the framework of Spanish, European, and USA research and development programs. These included three generations of vision chips, analog front-ends for XDSL MoDems, ADCs for wireless communications, ADCs for automotive sensors, chaotic signals generators, and complete MoDems for power-line communications. His research embraces smart imagers, vision chips, and biomedical circuits, always with an emphasis on system integration. He is on the committee of several international journals and conferences. He has chaired several international IEEE and SPIE conferences.

Design of Event-Driven SPAD-Based 3D Image Sensors

Rubén Gómez-Merchán,
Juan Antonio Leñero Bardallo and
Ángel Rodríguez-Vázquez

CRC Press
Taylor & Francis Group
Boca Raton London New York

CRC Press is an imprint of the
Taylor & Francis Group, an **informa** business

Designed cover image: © 2025 Shutterstock, atdigit

First edition published 2025
by CRC Press
2385 NW Executive Center Drive, Suite 320, Boca Raton FL 33431

and by CRC Press
4 Park Square, Milton Park, Abingdon, Oxon, OX14 4RN

CRC Press is an imprint of Taylor & Francis Group, LLC

Library of Congress Cataloging-in-Publication Data
Names: Leñero Bardallo, Juan Antonio, author. | 1 Gómez-Merchán, Rubén, aut |
1 Rodríguez-Vázquez, Ángel, aut
Title: Design of Event-Driven SPAD-Based 3D Image Sensors /
by Juan Antonio Leñero Bardallo, Rubén Gómez-Merchán and Ángel Rodríguez-Vázquez
Description: First edit | 1 Boca Raton, FL : CRC Press, 2 | Includes bibliographical references
Identifiers: LCCN 2024019471 (print) | LCCN 202401 (ebook) |
ISBN 9781032513539 (hbk) | ISBN 9781032547817 (pbk) | ISBN 9781003427490 (ebk)
Subjects:
Classification: LCC TK6592.O6 L46 2025 (print) | LCC TK6592.O6 (ebook) |
DDC 621.3848–dc23/eng/20241029
LC record available at https://lccn.loc.gov/2024019471
LC ebook record available at https://lccn.loc.gov/2024019472

ISBN: 978-1-032-51353-9 (hbk)
ISBN: 978-1-032-54781-7 (pbk)
ISBN: 978-1-003-42749-0 (ebk)

DOI: 10.1201/9781003427490

Typeset in Times
by Newgen Publishing UK

Contents

Preface/abstract

This book focuses on conceiving, developing, and implementing a novel vision sensor where *Single-Photon Avalanche Diodes* (SPADs) and event-driven architecture converge in a scalable architecture. The departure from conventional frame-based architectures represents a paradigm shift in SPAD-based sensors, aiming to alleviate the prevalent challenge of handling and processing extensive data volumes. By exclusively transmitting meaningful data, it eases storage and processing requirements, an advantageous characteristic for applications such as augmented reality and autonomous driving. The proposed sensor embeds 2D and 3D imaging capabilities. Also, its operation principle aligns with the paradigm of dynamic vision by introducing a discrete version of this paradigm that fits the operation of SPADs and supports motion detection and reduced data transmission.

The research begins by introducing the concept of an Event-Driven camera system specifically designed and tailored for *Light Detection And Ranging* (LiDAR) applications. This camera system encompasses the optical emitter, receiver (vision sensor), and required auxiliary circuitry for an autonomous operation. It directly processes information through events received from the vision sensor. Among its various functionalities, the camera system can dynamically adjust the sensitivity of individual pixels based on their absolute intensity values and detect intensity variations by analyzing the temporal information conveyed through these events.

The discrete arrival of photons presents challenges in devising circuits for motion detection. It motivates introducing the concept of a *discrete dynamic vision sensor*, whose behavior and metrics are analyzed in the book. Theoretical models proposed in this study are evaluated using experimental data, showcasing a methodology for *true-event-driven* dynamic vision with single-photon detectors that qualify for a digital implementation whose behavior would only be limited by photon shot noise.

The book then explores the implementation of the integrated circuit housing the vision sensor, elaborating on the pixel concept and its diverse implementations. A detailed discussion covers their respective advantages and drawbacks and suggests enhancements to augment and broaden the functionalities of the pixel.

Validation of the book proposals encompasses a bottom-up characterization of the system, covering the evaluation of SPAD devices until electrical verifications of the sensor. This comprehensive process culminates in validating the sensor in controlled lab conditions and real-world scenarios. Particularly notable is its validation in an astronomy application, highlighting the sensor's event-driven nature that enables the extraction of information at the single-photon level from the occultation of Betelgeuse by asteroid Leona with microsecond resolution, an unprecedented event in the field. This substantial advancement surpasses the temporal resolution limitations of conventional cameras, which are limited by their frame rate.

In summary, this book describes a SPAD-based event-driven architecture that lays the foundation for future research in single-photon sensors. Its functionality and versatility are assessed through extensive experimental validation, setting the stage for prospective advancements, and demonstrating remarkable potential to revolutionize single-photon imaging technology.

Abbreviations

ADAS	Advanced Driver Assistance Systems
ADC	Analog-to-Digital Converter
AER	Address Event Representation
APD	Avalanche Photodiode
APP	After-Pulsing Probability
APS	Active Pixel Sensor
ASIC	Application-Specific Integrated Circuit
BJT	Bipolar Junction Transistor
BTBT	Band-to-Band Tunneling
CCD	Charge-Coupling Device
CDS	Correlated Double Sampling
CIS	CMOS Image Sensor
CMOS	Complementary Metal-Oxide-Semiconductor
CoM	Center of Mass
CPM	Control and Processing Module
CTAT	Complementary to Absolute Temperature
CW	Continuous Wave
LiDAR	Light Detection and Ranging
DCR	Dark Count Rate
DNL	Differential Non-Linearity
DNW	Deep N-Well
DR	Dynamic Range
dToF	Direct Time of Flight
DVS	Dynamic Vision Sensor
EEL	Edge-Emitting Laser
EMVA	European Machine Vision Association
FB	Frame Buffer
FPGA	Field Programmable Gate Array
FPN	Fixed-Pattern Noise
FR	Free Running
FSM	Finite State Machine
FWHM	Full Width at Half Maximum
HDR	High Dynamic Range
HVLDO	High-Voltage Low-Dropout Regulator
IC	Integrated Circuit
INL	Integral Non-Linearity
iToF	Indirect Time of Flight
IO	Input/Output
LDO	Low-Dropout Regulator
LED	Light-Emitting Diode
LSB	Least Significant Bit
MPE	Maximum Permissible Exposure

MSB	Most Significant Bit
NIR	Near Infra-Red
PCB	Printed Circuit Board
PDP	Photon-Detection Probability
PPD	Pinned Photodiode
PLL	Phase-Locked Loop
PMU	Power Management Unit
PSR	Power Supply Rejection
PSRR	Power-Supply-Rejection Ratio
PTAT	Proportional-to-Absolute Temperature
PVT	Process Voltage and Temperature
QI	Quanta Imaging
ROI	Region of Interest
SAER	SPAD-based with AER communication
SNR	Signal-to-Noise Ratio
SPAD	Single-Photon Avalanche Diode
SPI	Serial Periphery Interface
SDRAM	Synchronous Dynamic Random Access Memory
SRAM	Static Random Access Memory
SWIR	Short-Wave Infra-Red
TDC	Time-to-Digital Converter
TIA	Transimpedance Amplifier
ToF	Time-of-Flight
TNP	Time-to-N Photons
TVC	Time-to-Voltage Converter
USB	Universal Serial Bus
VCSEL	Vertical-Cavity Surface-Emitting Laser

Tables

Figures

1 Introduction and challenges

Since the dawn of humanity, the ancestors of *Homo sapiens* have attempted to represent reality as they perceive it. Whether through simple cave sketches or complex artistic representations, capturing the environment has defined a pervasive challenge for human beings. The invention of photography enabled accurate, inexpensive recording of static events, thus paving the way towards making the capture of personal memories universally available. Today, microelectronics permits us to capture extra high-quality still pictures and videos with relatively cheap, light, handheld apparatus. Practically everyone worldwide carries a device in their pocket capable of immortalizing the visual scene almost precisely how we perceive it. Also, advances in understanding optical phenomena and scenes have taken image acquisition technologies to unprecedented levels [Yole23a].

The interpretation of visual scenes goes beyond the mere acquisition of the raw data associated with images and image sequences. Of course, these data can be analyzed by using conventional digital signal processing architectures based on the Von Neumann paradigm. However, the interpretation task is lengthy and power-hungry for sensors with hundreds of megapixels, each with as many bits per data as possible [Rodr23]. The challenge to increase the processing capability and decrease the power consumption of this conventional approach has prompted researchers to explore image-sensing architectures focused on extracting information (shapes, number of objects, salient points, etc.) instead of just raw data. Thus, image sensors with embedded parallel processing capabilities, also called vision chips, have been explored for decades, in many cases trying to emulate the principles and even the structures observed in natural retinas [Koch94]. This represents a complete paradigm shift: rather than focusing on obtaining the maximum amount of scene data, the goal is to extract meaningful information from it. In these types of sensors, pixels may transmit information asynchronously, disregarding the concept of frames. Concepts like *visual microprocessors* [Rosk01], *Dynamic Vision Sensors (DVSs)* [Lich06], and others are representative of scientific advances that are ramping up to industrial products.

Parallel to these advances, the incorporation of scene depth into the data captured by image sensors has experienced significant growth over the last few years [Yole23b]. Just as humans can perceive the depth of a scene when interacting with reality, modern applications call for 3D image sensors able to capture intensity

DOI: 10.1201/9781003427490-1

maps together with object ranges as needed for augmented reality, autonomous driving, or medical imaging, among others. While initial approaches to 3D imaging relied on using multiple cameras to estimate depth through triangulation [Bosc01], current sensors aim to estimate depth by computing the *Time of Flight* (ToF) of emitted light [Blai04; Bosc01]. In fact, recent research has demonstrated the possibility of manufacturing devices compatible with standard *Complementary-Metal-Oxide-Semiconductor* (CMOS) processes that can produce a measurable electrical signal upon the arrival of a single photon: *Single Photon Avalanche Diodes* (SPADs) [Pere18]. Although this may seem like the pinnacle of image sensing technologies, there are still challenges to be addressed [Palu14]. When measuring the ToF of a pulsed laser, multiple measurements and even histogram computations are necessary to obtain the definitive result, depending on the method employed for the implementation. This entails a tremendous data flow, resulting in high power consumption and resource utilization.

Despite significant advances in vision chips, on the one hand, and ToF sensors, on the other hand, the convergence of SPAD-based sensors and asynchronous architectures still needs to be resolved, as the number of received photons exceeds the bandwidth of such sensors. This book focuses on architectures to achieve this convergence, discussing design considerations and current limitations when designing asynchronous sensors based on SPADs. The combination of theoretical studies and experimental results provides an overview of the scenarios that allow us to assess the potential development of this technology.

1.1 BASIC CONCEPTS AND DEFINITIONS

In the realm of digital imaging, specifically referring to the technique of capturing visual information in digital format, a diverse array of components contributes synergistically. Figure 1.1 illustrates the fundamental processing sequence within a conventional *image sensor*. Foremost, a transducer, commonly denoted as a photoreceptor, undertakes the pivotal task of converting incident luminous flux, i.e., photons, into an electrical signal, which can be in the form of charge, voltage, or current.

Subsequently, readout electronics assume the responsibility of quantifying this electrical signal, operating as the interface that connects the analog and digital domains. This digital manifestation, subsequently capable of manipulation, may be subject to a sequence of computational operations by a digital processor. As an

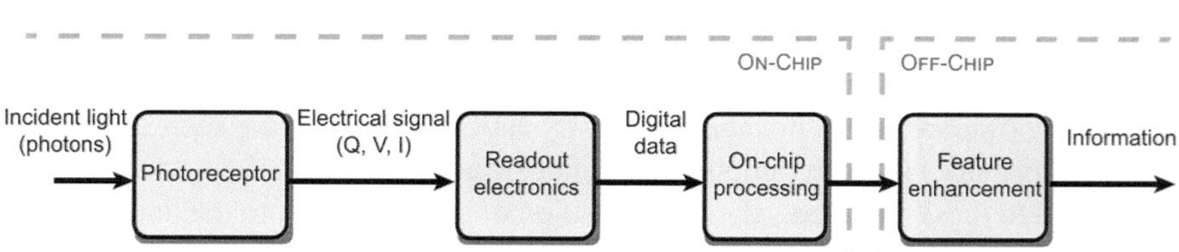

FIGURE 1.1 Fundamental processing sequence within a conventional image sensor. Feature enhancement is performed by the next element in the processing chain.

example, these operations might encompass functions such as color correction and noise reduction, among others.

Finally, the final output image that emerges from the image sensor may go through additional enhancement guided by an external processor. During this process, important features may be extracted from the content of the image.

Nevertheless, it is evident that this processing sequence might not be optimally efficient for various applications that extend beyond image representation and photography. The reason underlying this lies in the inefficiency inherent to conventional image sensors, which consume substantial energy and time to capture an image consisting of numerous pixels. From an application point of view, the information within the image might be related to only a limited subset of pixels. This inefficiency drives the exploration of an alternative sensor category known as *vision sensors*. Unlike their conventional counterparts, vision sensors carry out the crucial task of feature extraction, a task typically relegated to external processors in later stages. By doing so at an earlier stage, these sensors tailor their output to align with the specific demands of distinct applications.

Considering these factors, this section lays the foundation by providing essential definitions and concepts that are pivotal for delving into diverse vision technologies. By establishing this fundamental knowledge, we pave the way for an exploration of the various vision technologies and, in turn, enable a thorough understanding of the contributions made by this book.

1.1.1 Photoreceptors

Photoreceptors are a type of electronic device that converts light into electrical signals, similar to the way in which the photoreceptors of the human eye work. These devices have been widely studied in recent years and they are key components in modern imaging devices such as digital cameras [Dona21; Kris15].

The main component of every electronic image sensor is the same: a junction of two differently doped semiconductor materials, i.e., a photodiode. Depending on the application, this photodiode may have different physical configurations and operate in different modes, such as *linear, photovoltaic*, or *avalanche*. However, every implementation shares a common feature: it is a device capable of converting received photons into an electric charge.

Silicon-based photoreceptors offer many advantages, such as high sensitivity, high responsivity, high gain, low noise level, and, most importantly, they are cheap and easy to manufacture. Modern *Integrated Circuit* (IC) technologies mostly rely on CMOS processes. This allows for the inclusion of smart processing features at the sensor level.

However, understanding these devices is essential when designing such systems. Any type of electronic noise or manufacturing defect can introduce errors when measuring scene illumination. This is not only important in consumer electronics, where the ultimate goal is to obtain an image, but also in machine vision applications, which have additional requirements such as high dynamic range, artifact-free images, low noise levels, and linearity, to name a few.

Linear-mode Photodiodes

A basic photodiode consists of a p-n junction, where the p-type region (*anode*) is made of a material with an excess of holes (positive charges), and the n-type region (*cathode*) is made of a material with an excess of electrons (negative charges). When the photodiode is reverse biased, i.e., the voltage at the cathode is higher than at the anode, a depletion region is formed in the device as a result of a diffusion process.

Figure 1.2 depicts the band diagram of a reverse-biased photodiode, where essential energy levels are represented [Yadi04]. The Fermi energy level, E_f, defines the energy where electrons are equally likely to be occupied in the electron and conduction bands. The conduction band energy, E_c, represents the minimum energy level at which electrons achieve enough energy to move freely, contributing to electrical conduction. Conversely, the valence band energy, E_v, designates the maximum energy level for electrons that are tightly bound to atoms and do not contribute to electrical conduction.

The energy levels are influenced by a factor qV_b, where q is the electron charge and V_b is the reverse bias voltage. This factor causes energy levels to bend, creating an electric field ξ in the depletion region. Therefore, when a photon is absorbed and generates an electron-hole pair within the depletion region, it triggers the movement of these photo-generated carriers, allowing the accumulation or flow of charge. Figure 1.3 shows the I-V static characteristics of a p-n diode and defines different regions that can be used for photo-detection. Conventional photodiodes are biased in the so-called linear region, where the photodiode is reverse biased and, hence, currents produced by impinging photons are noticeable versus intrinsic currents due to thermal generation and other second-order phenomena [Dona21; Yadi04]. In this region the photocurrent, I_{ph}, is roughly proportional to the power of the incoming light, P_{opt}; i.e., to the number of impinging photons per unit time:

$$I_{ph} \simeq qQ_E \frac{\lambda}{hc} P_{opt} \tag{1.1}$$

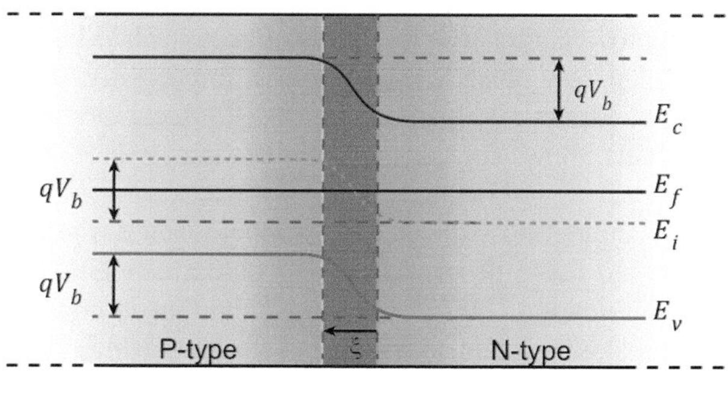

FIGURE 1.2 Band diagram of a p-n junction, showcasing the electric field at the depletion region.

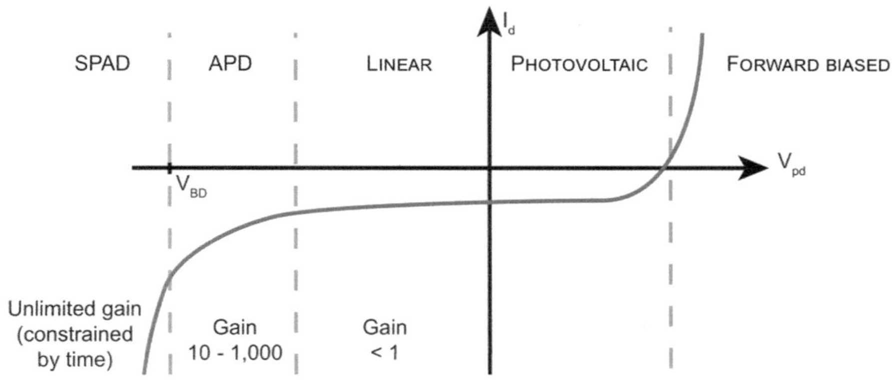

FIGURE 1.3 I-V characteristic of a photodiode, including the breakdown region.

FIGURE 1.4 Photodiode working in the a) accumulation mode, and b) current mode.

where Q_E is the quantum efficiency (ratio of photo-generated carriers to incident photons), h is the Planck's constant, c is the speed of light, and λ is the wavelength of the light. Note that the photocurrent may result from the contribution of multiple wavelengths, which are combined.

Although the photocurrent can be directly used as the signal vector, these photodiodes commonly operate in the *accumulation mode*, thereby integrating a charge that is proportional to the integration time. This strategy increases the signal level and provides an extra degree of freedom for extending the dynamic range and other features [Varg15]. Figure 1.4 (a) shows a photodiode working in the accumulation or integration mode. In this circuit, the photodiode is reset to V_b and then left floating. While holes find a low-impedance path to ground, electrons accumulate in the capacitor at the cathode. Therefore, if we define an integration time T_{int}, the output voltage V_o can be defined as:

$$V_o = V_b - I_{ph} \frac{T_{int}}{C_{ph}} \tag{1.2}$$

where C_{ph} is the equivalent in-parallel capacitance.

On the other hand, Figure 1.4 (b) depicts a basic circuit that measures the photocurrent, where a transimpedance amplifier sets the voltage of the photodiode and converts the current flowing through the photodiode, I_{ph}, to a voltage:

$$V_o = V_b - I_{ph} R \tag{1.3}$$

Note that the resistor in Figure 1.4 (b) can be replaced by a capacitor. In such scenario the photocurrent would be integrated in the capacitor, while keeping the photodiode at a constant bias, avoiding second order effects related to its operating point. The integration operation mode has become popular since adjusting T_{int} allows increasing the gain in equation (1.2), being the most common device in most image sensors. However, these devices and their associated electronics add a noise contribution to the signal, which hinders the operation at a single-photon level. Some studies have demonstrated that a noise lower than 0.3 e- rms is required to discriminate single photons in a linear-mode device [Tera12], although the model in [Foss13] suggests that a noise level of 0.15 e- rms is required for a proper operation.

Photodiodes are optimized for low-noise operation in the accumulation mode. Figure 1.5 depicts the cross section of a *Pinned Photodiode* (PPD) [Foss14]. PPDs feature a diode technology with a thin p+ layer, known as the pinning layer, which buries the diode. This configuration offers multiple benefits: firstly, it isolates the depletion region from recombination-generation centers on the silicon surface, thereby reducing dark current (non-photon absorption-related current). Secondly, the PPD operates at a voltage denoted as the pinned voltage, achieving full depletion and enabling true charge transfer through a transfer gate. Charge transfer to a floating diffusion with lower capacitance than the PPD enhances conversion gain, effectively reducing input-referred noise.

Photovoltaic-mode photodiodes

As an alternative to integrating charge-based photodiodes, photovoltaic-mode devices feature continuous operation, inherent logarithmic compression, and the possibility to combine image acquisition and energy harvesting [Gome23a; Gome23b; Ni11].

Figure 1.6 (a) shows the I-V relationship of a photodiode and its dependency on the illumination level, defining three different regions depending on the sign of the voltage and current:

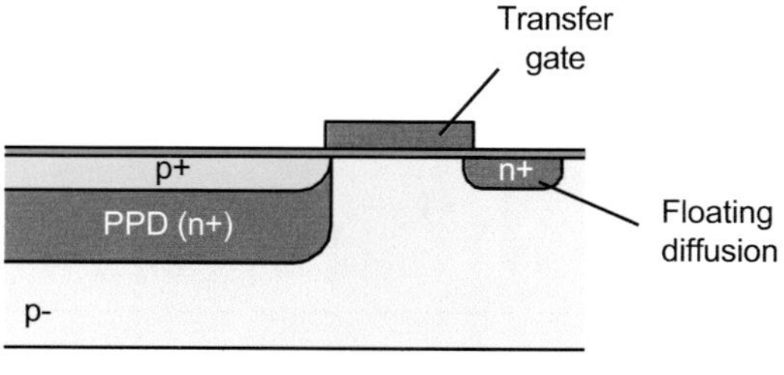

FIGURE 1.5 Cross section of a Pinned Photodiode.

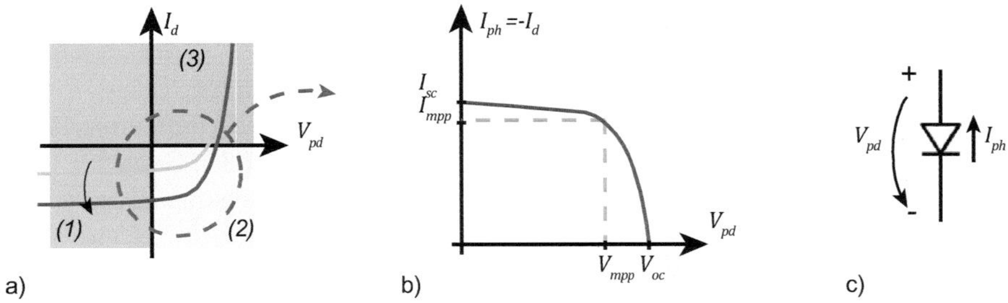

FIGURE 1.6 a) I-V characteristic of a photodiode representing the (1) reverse-biased, (2) photovoltaic, and (3) forward-biased region. b) I-V characteristic of a solar cell. c) Current and voltage polarity of a photodiode operating in the photovoltaic mode.

1. **Reverse-biased region.** The photodiode consumes energy and presents a quasi-linear dependance on the illumination, as mentioned previously.
2. **Photovoltaic region.** In this region, the photodiode transforms incident light into usable energy that can be used to supply circuits.
3. **Forward-biased region.** The photodiode acts as a regular diode, providing a low-impedance path from anode to cathode. The dependence on light is masked by the intrinsic diode current.

When photodiodes are employed as solar cells or photovoltaic modules, it is usual to represent the I-V curve as in Figure 1.6 (b), following the notation defined in Figure 1.6 (c), where the current is assumed to flow from the cathode to the anode.

Let us consider a scenario where the cathode of the photodiode is grounded, and the anode is left unconnected. In this configuration, the photocurrent charges the inherent capacitance of the photodiode. As a result, a voltage builds up until the photocurrent is balanced by the forward-bias current. The following parameters are of interest when operating within the photovoltaic region:

- ***Open-circuit voltage, V_{oc}.*** This is the equilibrium voltage where the photocurrent is equilibrated by the forward-biased current.
- ***Short-circuit current, I_{sc}.*** This current occurs at $V_{pd} = 0\ V$ and can be approximated by the reverse-biased current.
- **Maximum power point.** Represented by I_{mpp} and V_{mpp}, this point denotes the highest power achievable from the solar cell.

From an imaging point of view, photodiodes serve a dual purpose: they can harvest energy during idle states [Choi12; Gome22; Gome23a; Park18] and encode illumination levels using V_{oc} [Gome21; Gome23c; Ni11]. This voltage can be expressed as:

$$V_{oc} = \eta U_t \ln\left(\frac{I_{ph}}{I_s}\right) \tag{1.4}$$

where η is the ideality factor, U_t is the thermal voltage, and I_s is the reverse-biased saturation current.

However, these devices pose several disadvantages. Foremost, equation (1.4) indicates that V_{oc} presents a strong temperature dependence. Furthermore, the signals generated in these devices are typically low [Cili14; Ferr19; Gome20]. This, combined with the slow speed of photovoltaic-mode devices when transitioning from bright to dark due to a drift current induced by a weak and decreasing electric field, poses a clear disadvantage when integrating them into an image sensor. This is especially significant in situations where temperature gradients and rapid scene variations are in place.

Overall, photodiodes in photovoltaic mode are an important technology for converting light energy into electrical energy, with a wide range of applications in solar power generation and energy harvesting. Ongoing research in this area is aimed at improving the efficiency, stability, and cost-effectiveness of these devices, making them increasingly viable for large-scale applications. However, these devices still need to be fully qualified for industrial imaging applications.

Avalanche and single-photon avalanche photodiodes

Avalanche Photodiodes (APDs) and SPADs are also made of a p-type semiconductor layer and an n-type semiconductor layer, but in this case, the bias point varies from linear-mode and photovoltaic-mode photodiodes.

The I-V characteristic of a photodiode shown in Figure 1.3 also depicts the breakdown region, which accounts for the avalanche effect of photodiodes. As the reverse bias voltage applied to a photodiode increases, so does the corresponding electric field, increasing the kinetic energy of charge carriers. Once this voltage exceeds the breakdown voltage, V_{BD}, the phenomenon of impact ionization comes into play. During impact ionization, charge carriers can transfer their energy by generating additional charge carriers, leading to a cascading effect similar to a chain reaction.

Impact ionization is of special interest in sensing applications due to its potential for enhancing the gain of photodiodes. While linear-mode photodiodes exhibit a gain (i.e., generated carriers per incident photons) slightly below one, APDs operate near V_{BD}, in a region where there is a balance between the generation and quenching of avalanches. As a result, the electron-hole pair generated during photon detection can trigger the creation of additional carriers, increasing the gain. However, APDs pose challenges in terms of stabilization, non-linear gain, and exhibit an ill-defined gain, thereby imposing several constraints when integrating them into high-resolution arrays [Seit13].

SPADs operate in what is known as the Geiger mode. SPADs are stably biased above the breakdown voltage. In this configuration, when a single photon is detected by the SPAD, the device transitions into a state of breakdown, resulting in a large current flowing through the device that can be easily measured. This enables the detection of signals down to the single-photon level. However, the breakdown state is destructive, requiring extinguishing the avalanche after its occurrence. Figure 1.7 (a) shows a SPAD along with the quenching circuitry responsible for extinguishing the avalanche. Figure 1.7 (b) illustrates the operation of the SPAD. Initially (1), the

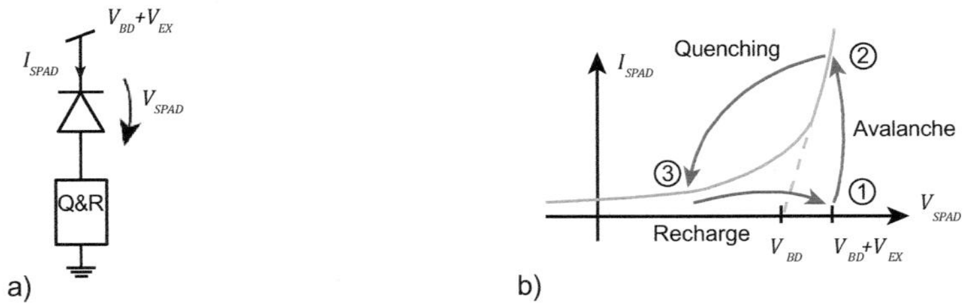

FIGURE 1.7 a) Fundamental SPAD Configuration with Quenching and Recharge Circuitry. b) Illustration of SPAD Operation Diagram.

SPAD voltage begins at $V_{BD} + V_{ex}$, with V_{ex} denoting the excess voltage or the increase from V_{BD}. Upon photon detection, the SPAD enters the avalanche state, indicating the detection event (2). Subsequently, the quenching circuitry identifies this current and reduces the voltage of the SPAD below V_{BD} (3), effectively terminating the avalanche. Once the avalanche is adequately quenched, the SPAD is recharged to its initial voltage, a process facilitated by either the quenching circuitry or supplementary circuitry.

These devices have been a game-changer in various sensing applications, as they have unlocked previously unthinkable functionalities. SPADs yield measurable signals with sub-nanosecond accuracy, enabling their correlation with the detection of individual photons. This along with the integration of miniaturized arrays and processing circuitry on a single IC enables a multitude of applications, particularly in the realm of low-light sensing and photon correlation. Examples encompass positron emission tomography [Brag14; Tetr15], depth estimation [Gyon23; Mori20a; Rocc20; Zhan19a], or fluorescence lifetime imaging microscopy [Ulku19], among others.

However, these devices are far from being ideal. In addition to avalanches caused by the absorption of a photon, spurious avalanches—also known as *dark counts*—limit the device operation. The *Dark Count Rate* (DCR) imposes a lower limit on the arrival photon rate that we can measure. This, combined with photons from the ambient light itself, complicates the detection of correlated photons, imposing the need for oversampling and building histograms to diminish the contribution of uncorrelated photons. This distinctive attribute imposes additional constraints at both the circuit and architectural levels, as it requires the handling of substantial volumes of data. Furthermore, much like linear-mode photodiodes exhibit a quantum efficiency lower than one, SPADs do not capture every individual photon that interacts with the silicon crystal but do present a *Photodetection Probability* (PDP) contingent on the wavelength of the incident photon.

Silicon-based SPADs are limited to the visible and *Near-Infrared* (NIR) spectrum. They typically exhibit PDP peaks, often situated at blue or green wavelengths, approaching up to 50% in most cases [Gram22; Leit13; Nicl07; Veer15]. While the NIR holds potential for active illumination systems due to its reduced harm to the human eye, the PDP can drop well below 5%.

In addition to all of this, the avalanche voltages of photodiodes in standard CMOS technologies are relatively high (15-30 V [Gram22; Leit13; Nicl07; Veer15]), introducing high electric fields which, combined with the need to include isolation structures and guard rings, constrain the miniaturization of these devices.

Despite the inherent challenges of SPAD technology, it is progressively gathering interest, capturing the focus of both researchers and cutting-edge companies such as Sony [Kuma21; Ogi21]. Nonetheless, it still faces several challenges that demand attention at the device and circuit level. Thus, the field presents an exceptional opportunity to develop innovative sensors for the future [Cecc21; Cusi22].

1.1.2 Image vs. Vision sensors

While the terms *image sensor* and *vision* sensor are at times used interchangeably and these concepts can sometimes be open to interpretation, we can define image sensors as devices that are engineered with the objective of capturing scenes in a manner that closely emulates the human visual experience. This entails striving for a representation that preserves the highest level of scene fidelity, detail, and image quality—similar to what human retina does concerning the capture of fine image details although ignoring that retinas do not capture images but also pre-process them [Rosk06; Zhao06].

Indeed, when we examine the human visual system for comparison, it becomes evident that our own eyes do not directly render the images we perceive. Rather, the human *retina* comprises layers of cells that capture different aspects of incoming light, such as color opponency, intensity, and motion [Purv17]. The brain then processes and integrates data from these diverse cortical layers to create the entire visual experience that we perceive as *seeing*. This multi-layered approach allows the human visual system to efficiently process and interpret visual information, a paradigm that inspires the development of advanced imaging technologies.

Thus, we may define a vision sensor as a device that is strategically designed to extract specific features or relevant information from a scene while bypassing the resource-intensive processes of full-frame, high-resolution image processing [Belb10; Zara11]. By targeting precise feature extraction, vision sensors act as a *front-end* for data reduction, enabling the removal of power-hungry and time-consuming tasks associated with processing extensive full-frame images [Rodr18]. This approach not only conserves computational resources but also reduces the volume of transmitted data and the storage requirements associated with the processing pipeline.

In essence, the distinction between image sensors and vision sensors underscores the goals of each technology. Image sensors aim to capture optical scenes with full level of details, striving for faithful scene representation. In contrast, vision sensors optimize the data collection process, emphasizing feature extraction to deliver efficient, targeted insights from the visual scene. Through these distinct approaches, both technologies contribute significantly to the broader landscape of visual data capture and interpretation.

To accurately classify image and vision sensors based on the previous definition, it becomes imperative to establish a framework that encompasses key concepts essential to their functionality. This involves defining parameters such as *image quality, image artifacts*, or *feature extraction.*

Image quality and imaging metrics

Exploring the concept of image quality is complex, involving various factors and considerations [Naka06]. While establishing a definitive definition can be challenging due to the subjective nature of human perception and the diverse uses of images, it is possible to identify key metrics that contribute to the overall quality of an image [EMVA10]. These metrics serve as important criteria for evaluating how faithful, visually appealing, and useful a captured image is.

- **Spatial resolution** refers to the ability of an imaging system to spatially differentiate between fine details within an image. The sensing region of a digital image sensor is divided into discrete units called *pixels.* In this way, the pixel serves as the fundamental unit of the sensor, representing the smallest measurable piece of information. Therefore, spatial resolution dictates how closely individual elements or features within an image can be discerned. Higher spatial resolution results in finer details and greater clarity, while lower resolution may lead to a loss of intricate features, as shown in Figure 1.8 (a) and (b).
- ***Dynamic Range*** (DR) defines the capacity of the image sensor to capture a broad spectrum of light intensity levels. A *High-Dynamic-Range* (HDR) image captures detailed information in both bright and dark areas of the picture, keeping details that might be lost in non-HDR or low DR images, which can become overly bright in well-lit areas (see Figure 1.8 (c)) or too dark in dark spots. DR is commonly discussed in terms of *intra-frame c*hanges and contrasts, showcasing the aptitude of the sensor for preserving details across varying light levels within a single image. However, HDR techniques can go further by combining multiple frames captured at different exposure times or locally adjusting the exposure time, expanding the range of captured light intensities.
- **Noise**. Similar to how audio signals can be disrupted by noise—even leading to useless signals—image sensors also exhibit some degree of noise, as represented in Figure 1.8 (d) [Sego17]. In this context, two distinct categories can be defined: *Fixed-Pattern Noise* (FPN) and temporal noise. FPN arises from systematic discrepancies among pixels and readout electronics due to component mismatches within the sensor. However, as this noise follows a systematic pattern, circuit techniques can be employed to mitigate FPN [Enz96], resulting in notably low levels. In contrast, temporal noise is related to fluctuations over time in the image, unrelated to scene variations. Minimizing noise has posed a significant challenge over the past decade, as it holds vital importance for both imaging and vision tasks. It is typically quantified in terms of electrons (e-), given that most image sensors employ photodiodes operating in the accumulation mode.
- **Color accuracy** refers to the ability of the sensor to faithfully reproduce colors as they appear in the real world. It entails capturing and conveying colors with precision, ensuring that the hues and tones captured by the sensor closely match

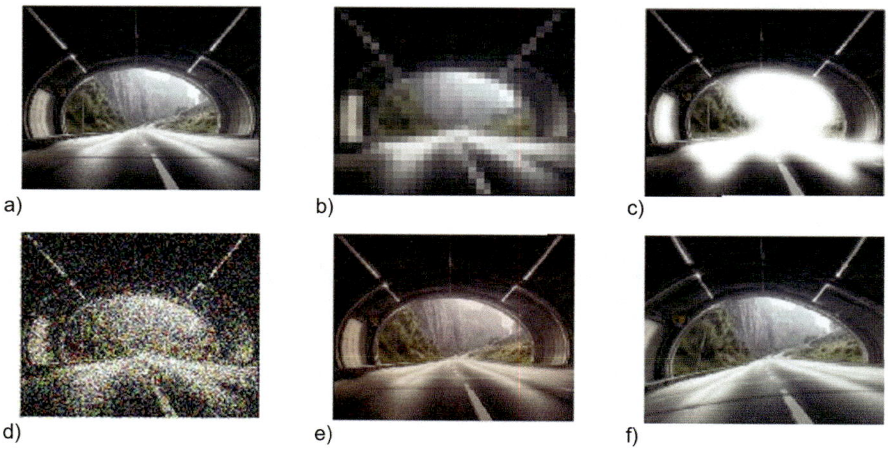

FIGURE 1.8 a) Example of HDR Image. b) Low-resolution image. c) Low DR image. d) Noisy image. e) Image with incorrect color correction. f) Distorted image.

those perceived by the human eye under the same lighting conditions. Accurate color reproduction is imperative in fields like product photography or consumer electronics, where faithful representation is sought by the users. Figure 1.8 (e) illustrates the effect of incorrect color correction in image sensors.

- **Distortion**. While not an intrinsic metric of the image sensor itself, distortion is a crucial aspect to consider when characterizing the entire imaging system or camera. Lenses can distort the intended representation of objects in an image and result in perceptual irregularities. Figure 1.8 (f) shows an example of distorted image. Additionally, either non-linear or non-uniform behavior of photoreceptors and readout circuits may contribute to signal distortion and therefore, to photo-response non-uniformity.

In the context of quantifying and standardizing these elements, the *European Machine Vision Association* (EMVA) provides a vital contribution through its established standards [EMVA10]. EMVA standards define metrics and methodologies to assess image quality parameters in a consistent and reproducible manner across various imaging technologies. These standards enable industry professionals to objectively evaluate and compare image quality attributes, fostering a common language for technical specifications and performance assessment. As a result, the EMVA standards play a pivotal role in bridging the gap between the abstract concept of image quality and its concrete evaluation, thereby enhancing the accuracy and utility of image sensor technologies in real-world applications. However, when the goal of a sensor is not to obtain an image but rather to extract information from the scene, the scenario may become complicated. Existing standards do not cover the characterization of sensors designed for such vision-oriented tasks.

In summary, while the concept of image quality remains complex, the key elements of spatial resolution, dynamic range, noise, color accuracy, and distortion collectively serve as the foundational pillars that assess and evaluate the quality of captured images across diverse applications. This evaluation, supported by EMVA and emerging standards, facilitates the comparison of different imaging and vision technologies.

Image artifacts

Within the field of CMOS image sensors, the quest for flawless image acquisition may be hindered by the appearance of image artifacts, i.e., unwanted anomalies that disrupt the accuracy of visual representation. A thorough understanding of these artifacts is crucial for discerning their underlying causes and defining effective methods to improve overall image quality. Among most common image artifacts, we can highlight:

- **Blooming**: When an overexposed pixel cannot accumulate more charge, i.e., they become saturated, excessive charge is absorbed by neighboring pixels and blooming occurs. This artifact is characterized by a halo of light that spills into adjacent regions. Blooming is most noticeable in scenes featuring bright spots, such as the sun. To address blooming, careful exposure control and the implementation of anti-blooming mechanisms can be employed.
- **Color artifacts**: Deviations from accurate color representation give rise to color artifacts, including phenomena like color fringing, false colors, and shifts in color balance. Lens imperfections, sensor design, and non-optimal lighting contribute to these discrepancies, impacting faithful color depiction. Techniques like color correction algorithms and improved sensor design help minimize color artifacts and enhance color accuracy.
- **Motion blur:** Rapid movement during exposure intervals causes motion blur, resulting in blurred or distorted object representation. This artifact becomes evident in low-light conditions, where longer exposures times are required. Motion blur diminishes image sharpness and overall clarity.
- **Aliasing**: When scene details contain higher frequencies than the spatial sampling capability of the sensor, i.e., the sensor resolution, aliasing emerges. This usually manifests as a moiré pattern [Sido02], distorting the image. Using anti-aliasing filters and post-processing helps mitigate aliasing artifacts.

A solid understanding of the origins and manifestations of these image artifacts empowers image sensor engineers to implement effective strategies, enhancing image quality and ensuring coherence between scenes and output images. Understanding the nature of these artifacts also enables consideration of their effects in designing alternative image sensor architectures.

Feature extraction

The field of feature extraction plays a pivotal role in vision systems, aligning with the principles outlined in previous discussions. Unlike the conventional approach that focuses on exhaustive data processing through number crunching using Von Neumann architectures, feature extraction targets the identification and extraction of pertinent information from the scene [Lopi11; Suar17; Zara11]. Traditionally, this stage has been relegated to off-chip processing by external processors, which, while effective, entails significant wastage of energy, time, memory, and processing resources. The challenge lies in the fact that such traditional methods process data indiscriminately, including both meaningful and redundant pixels. Indeed, this external processing may become the bottleneck of the entire system.

This inefficiency has encouraged the evolution of two distinct paradigms in imaging systems: *data-centric* and *information-centric*. In the *data-centric* realm, represented by image sensors, the emphasis remains on extracting information from all pixels in a manner that maximizes faithfulness to the scene. On the other hand, *information-centric* systems, represented by vision sensors, integrate early-stage vision tasks directly into the sensor. This paradigmatic shift allows for focal-plane processing, reducing the need for processing redundant data in a further stage. In an *information-centric* framework, the sensor output is tailored to provide precisely the data required for decision-making, eliminating the need to process meaningless data [Rodr18].

Figure 1.9 (a) visually outlines the different steps within the processing chain of an image sensor, including demosaicing, exposure control, and noise removal. In contrast, Figure 1.9 (b) depicts the processing steps adopted by a vision sensor, which also may include noise removal, in addition to edge detection, feature extraction, and identification task [Belb10]. It is important to note that the output of a vision sensor does not necessarily manifest as a high-resolution frame. As an example, the output might comprise a list of recognized objects along with their respective boundaries. These processing chains are illustrative examples and can incorporate more sophisticated stages to fulfill the specific requirements of the system.

The concept of feature extraction, as previously explained, gains significant interest when considered within the context of focal-plane processing—a visionary

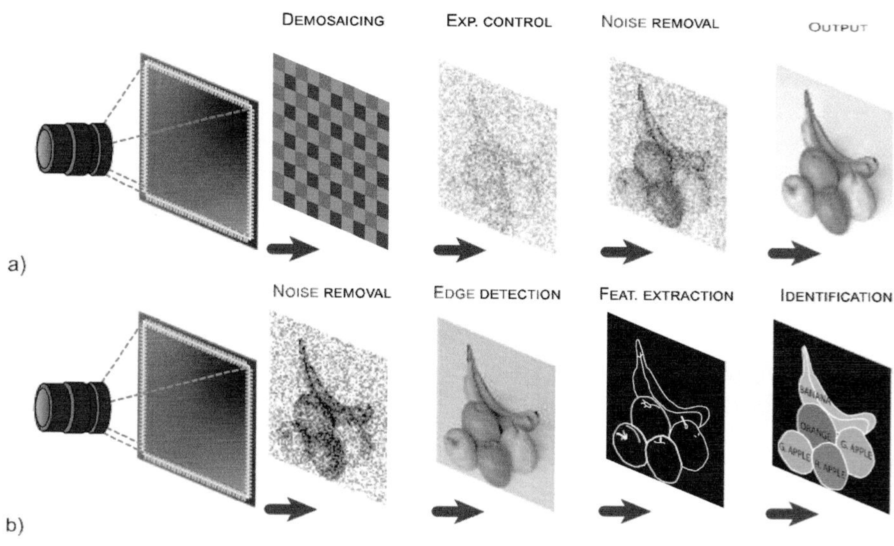

FIGURE 1.9 a) Different steps within the processing chain of an image sensor. b) Steps within the processing chain of a vision sensor.

shift that recalibrates the very foundation of the conventional processing chain. In this novel paradigm, pixels transcend their traditional role as passive data repositories to become independent entities including smart functionalities.

The archetypal processing chain illustrated in Figure 1.9 assumes a sequential readout of individual pixels, with each pixel contributing its portion to the mosaic of the complete image. Focal-plane processing, however, redefines this sequence by looking beyond images. Each pixel becomes a discrete processing element with computational ability. Pixels may no longer operate in isolation, instead they may exchange information among their neighbors. Thus, information may no longer be related to individual pixels but to a set of pixels.

The transition is from a mere accumulation of pixel signals to the establishment of a distributed framework for signal interpretation. In vision sensors, clusters of pixels may interact to extract context and meaning. This approach introduces an innovative way of understanding scenes, where every pixel has a unique role, working together to create a unified understanding of the scene.

This paradigm also enables sensors where pixels are the main characters that orchestrate the operation: focal-plane processors [Zara11] and *event-driven* architectures [Lene18]. These architectures can give rise to a significant transformation in data handling. Rather than scanning and processing each pixel indiscriminately, pixels autonomously request access to the readout channel only when an *event* occurs. These architectures aim to conserve resources, dedicating computational energy and resources exclusively toward extracting information of actual relevance.

In essence, focal-plane processing leads to a departure from the conventional paradigm. The shift from individual pixels data to useful information aspires to improve efficiency, data generation, and speed of vision systems. As the field advances, focal-plane processing demonstrates its potential in vision sensors that overcomes existing limitations [Rodr18], aligning a trajectory toward a new frontier in vision technologies.

1.1.3 LiDAR Systems

An example of a vision system of particular interest within the scope of this book is the *Light Detection And Ranging* (LiDAR) system [Li20; Peng23; Weit06; Yang22]. LiDAR systems hold immense significance due to their wide array of applications, ranging from autonomous vehicles and environmental monitoring to topographic mapping and robotics.

The fundamental goal of LiDAR systems is twofold: to acquire light intensity maps, i.e., 2D images, and to determine the depth to the scene. To accurately measure distances, the system controls a correlated source of light and measures the time it takes for the light to return after reflecting off an object, i.e., the ToF. Figure 1.18 illustrates this process, where the emitted light is a light pulse and t_{tof} is the ToF. Assuming that light is traveling through the air, we can approximate its speed to the speed of light in vacuum, c, thus calculating d as:

$$d = \frac{c}{t_{tof}} \times 2 \tag{1.5}$$

LiDAR systems can be classified based on several key criteria that determine their operational principles and capabilities. One primary criterion involves the method used to compute ToF. In one approach, ToF can be derived through *indirect ToF* methods (iToF), where the phase shift of the return signal is measured rather than the direct time taken for the light pulse to travel to the target and back. Within this category, there exists a distinction between *Continuous-Wave* (CW-iToF) modulation, where the emitted signal is a sinusoidal wave, and *Pulsed* (P-iToF) modulation, which utilizes a pulse train for the signal.

Alternatively, with the advent of SPADs in CMOS processes and their remarkable time resolution, certain LiDAR systems now demonstrate the capability to directly measure the time interval between the emission and reception of a laser pulse, resulting in a *direct ToF* (dToF) approach. The dToF category can be further divided into two subcategories: those with moving parts, referred to as Scan LiDAR, and those without, known as Flash LiDAR. Figure 1.11 illustrates this classification.

Each category of LiDAR technology offers distinct advantages and disadvantages. Indirect methods can offer higher resolution and lower error at near distances, allowing for greater distance accuracy; however, accuracy degrades linearly with distance. CW is suitable for applications requiring moderate distance measurements, while pulsed modulation is usually more robust to background in addition to a lower computational complexity, at the expense of being more prone to distortion, as squared signals have more harmonics than sinewaves.

On the other hand, direct methods provide more straightforward measurements since dToF does not require complex calculations. However, multiple samples acquisition and even histograms building is required to compute ToF [Gyon22]. Scan LiDAR

FIGURE 1.10 Illustration of the operation of a LiDAR system.

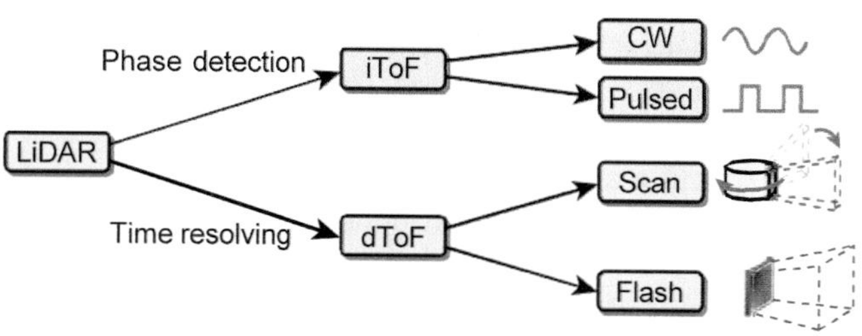

FIGURE 1.11 LiDAR Classification Based on ToF method, light waveform, and moving parts requirement.

TABLE 1.1
Comparison between iToF and dToF technologies.

Technology	Advantages	Disadvantages
Indirect Time-of-Flight	• Mature technology. • Compatible with traditional image sensing paradigm. • High fill factor and resolution.	• Range-resolution trade-off. • More sensitive to multi-path reflections. • Generally limited to range to tens of meters.
Direct Time-of-Flight	• High depth accuracy. • Longer depth range, independent of resolution. • Enables differentiation of semi-opaque surfaces and multi-path reflections.	• More complex implementation and data generation. • Limited fill factor. • Less power-efficient due to multiple laser exposures.

enables a wide field of view and detailed scene capture, but their implementation often comes with higher costs and larger form factors. This is mainly attributed to the inclusion of mechanical components, which can increase susceptibility to damage. In contrast, Flash LiDAR, which lacks moving parts, can achieve rapid data acquisition, and provide instant full-frame measurements. However, it might face challenges in maintaining high timing accuracy.

Ultimately, the choice of LiDAR technology depends on the specific application requirements and trade-offs between accuracy, speed, complexity, and cost. As LiDAR systems continue to evolve, these different categories offer a diverse range of solutions to address various real-world scenarios. Currently, iToF manifests as a more mature technology, resulting in a cheaper solution for depth estimation, however, academic institutions and cutting-edge companies are investing resources and effort on the development of dToF sensors. Table 1.1 summarizes the main advantages and disadvantages of each technology.

Indirect Time-of-Flight

As previously discussed, iToF is a prominent technique in LiDAR systems that operates based on the phase shift of modulated light waves to determine the distance to an object. This technique offers advantages such as flexibility, reasonable accuracy at near distances, and acceptable frame rates.

CW iToF operates by continuously modulating the emitted light with a sine wave, while a sensor measures the phase shift of the received light. This phase shift, represented as $\Delta\phi$, serves as a key parameter for distance estimation. Figure 1.12 illustrates the entire process. The distance, d, between the sensor and the target can be calculated using the equation [Hans13]:

$$d = c\frac{\Delta\phi}{2\pi f_m} \tag{1.6}$$

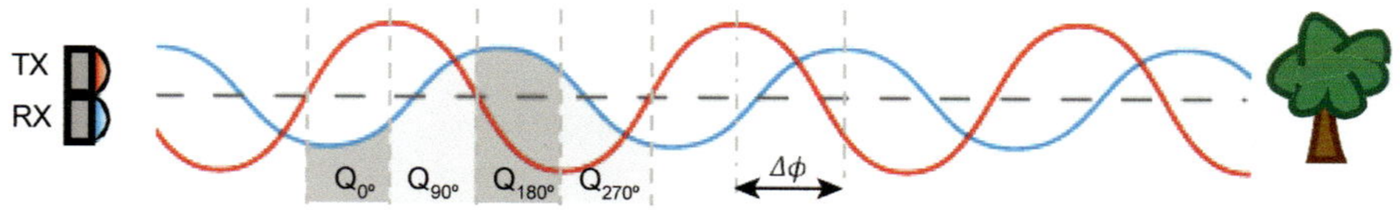

FIGURE 1.12 Illustration of the CW-iToF operation.

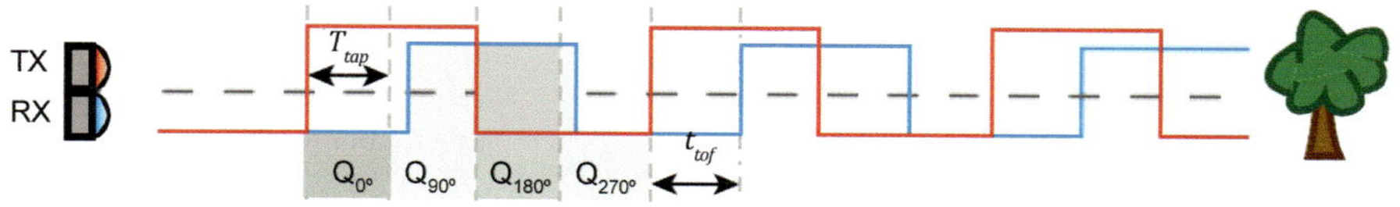

FIGURE 1.13 Illustration of the P-iToF operation.

where f_m represents the modulation frequency of the sine wave. To determine the phase shift $\Delta\phi$, a photodetector accumulates charge across four different phases, each shifted by 90°. The acquisition of these four phases may require more than one exposure, a topic that will be further discussed in subsequent sections. These samples obtained from the phase-shifted measurements are commonly referred to as *taps*. Given that Q_i represents the charge accumulated during phase i, the computation of $\Delta\phi$ can be expressed as [Hans13]:

$$\Delta\phi = \arctan\left(\frac{Q_{270^\circ} - Q_{90^\circ}}{Q_{0^\circ} - Q_{180^\circ}}\right) \tag{1.7}$$

From equations (1.7) and (1.8), it can be easily deduced that the demodulation process is not straightforward, on the one hand, and that the measurable distance is limited by the modulation frequency, on the other. In fact, longer distances may alias, leading to errors, as well in the case of multi-paths signals [Fuch10]. This can easily be solved by using multiple modulation frequencies, at the cost of increasing the complexity of the system and limiting the frame rate.

Conversely, P-iToF sensors adopt a square signal modulation for the emitted light, as depicted in Figure 1.13. This characteristic offers several advantages in terms of simplicity, since ToF can be computed as:

$$t_{tof} = \frac{T_{tap}}{2}\left(1 + \frac{Q_{180^\circ} - Q_{0^\circ}}{Q_{90^\circ} - Q_{270^\circ}}\right) \tag{1.8}$$

where T_{tap} is the width of the integration window of each tap. Notice that no need for any trigonometrical ratio is required.

When comparing CW-iToF and P-iToF techniques, it becomes evident that both methods demonstrate depth precision affected by factors such as the distance range, the intensity of the received light, and background illumination. CW-iToF exhibits distance-independent precision as it captures the entire reflected light, ensuring consistent *Signal-to-Noise Ratio* (SNR) across distances. Conversely, the precision of P-iToF is influenced by distance. In P-iToF, the signal collected at each phase shift varies with distance, resulting in variable SNRs and potential distance errors [Bell13].

Direct Time-of-Flight

The advent of single-photon detectors, such as SPADs, has paved the way for the development of dToF sensors. One of the primary advantages of dToF technology lies in its resolution, which remains independent of the range, unlike in the case of iToF sensors. This distinction arises from the method used to measure this time. In dToF, it is directly measured from a correlated light source, which means that the depth resolution is limited by the resolution of the circuit measuring the time, while the range does not necessarily affect this resolution.

To compute ToF, single-photon detectors are coupled with circuits known as *Time-to-Digital Converters* (TDCs). Figure 1.14 (a) provides an illustration of a TDC. Typically, it includes a signal to initiate the measurement process, synchronized with the light source, and a stop signal that freezes the output value of the TDC. In-pixel TDCs with time resolutions of below 40 ps have been reported [Hend19]. However, since not all detected photons necessarily originate from the light source, building a histogram is required to identify correlated photon arrivals.

While the most straightforward TDC architecture may involve digital oscillators or delay lines [Henz10], alternative solutions that slightly deviate from the concept in Figure 1.14 (a) can also be implemented. Figure 1.14 (b) demonstrates how a histogram can be built by defining different temporal windows associated with bins, each shifted by a specific delay relative to the light pulse, similar to a *Boxcar* average. This categorizes photon arrivals into different bins, directly constructing a histogram from photon counts. It is evident that the width of each bin, contingent on the generation of the temporal window, hardly reaches the resolution of high-performance TDCs. Nevertheless, sub-bin resolution can be achieved through interpolation [Gyon22].

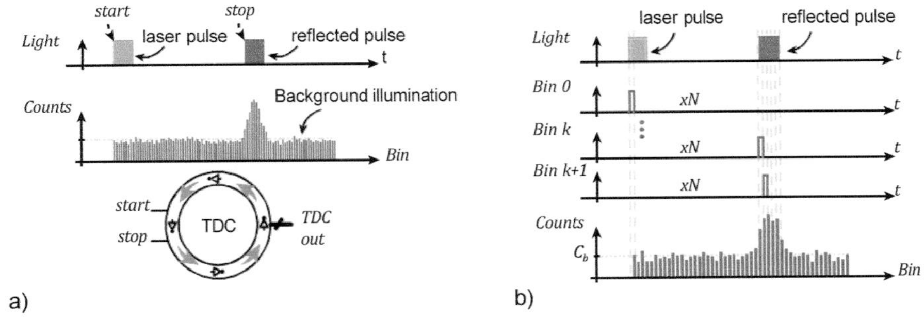

FIGURE 1.14 P-ToF histogram building using a) Time-resolved circuits (TDCs). b) Time-gating.

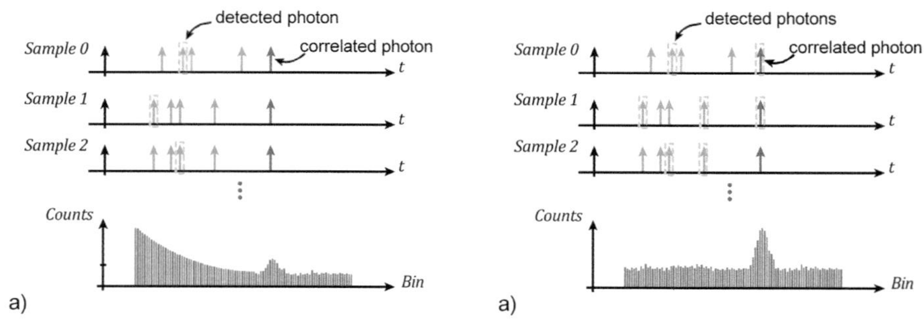

FIGURE 1.15 Detected photons and resulting histogram in a) single-event, b) multi-event TDC.

Essentially, the *Center of Mass* (CoM) of the histogram can be computed after subtracting background-illumination photon counts, C_b. Thus, ToF is then calculated as [Gyon22]:

$$t_{toj} = \frac{\sum_{i=0}^{n}(C_i - C_b)T_i}{\sum_{i=0}^{n}(C_i - C_b)} \tag{1.9}$$

where C_i and T_i are the photon count and time shift of bin i, respectively.

A critical aspect of the TDC employed in the dToF computation process is its capability to detect multiple photon events. TDCs equipped for this purpose are commonly referred to as multi-event TDCs [Henz10]. In cases where a TDC only records the timestamp of the first arriving photon, the histogram may suffer from pile-up under high background illumination conditions, as depicted in Figure 1.15. This arises because there is a high probability of a non-correlated photon triggering the TDC. On the contrary, if the TDC can retain the timestamps of multiple photon arrivals and the laser power is sufficiently high, pile-up distortion does not occur since correlated photons will also be detected.

1.2 CMOS IMAGE AND VISION SENSOR ARCHITECTURES

Over the past few decades, the evolution of *CMOS Image Sensors* (CISs) has been clearly noticeable. Unlike their conventional counterparts like *Charge Coupling Devices* (CCD), the incorporation of processing circuitry at the chip level has triggered a revolution enabling a diverse range of applications and giving rise to smart vision sensors capable of executing complex visual tasks [Belb10]. This fact has not only driven a significant transformation in imaging capabilities but has also paved the way for ToF techniques to be integrated into sensor architectures [Hans13].

In this section, we will embark on the exploration of the most prevalent CMOS image sensor architectures, from the pixel level to a system-level perspective. We will describe several architectures based on accumulation-mode photodiodes and SPAD technology.

Moreover, we will delve into the field of *event-driven* architectures, specifically designed to implement dynamic vision tasks with remarkable efficiency. These architectures aim to replicate some of basic behavior of human retinas concerning signal coding and reaction to changes [Rosk06], enabling tasks that go beyond conventional imaging scenarios. Finally, we will discuss most common implementations of iToF and dToF techniques described in Section 1.1.3.

1.2.1 Conventional APS image sensors

Active Pixel Sensors (APS) [Foss14; Foss97; Nobl68] represent a significant advancement in the field of CIS technology. In contrast to passive-pixel sensors, APS embed amplification directly at the pixel level, making them inherently more suitable for high-resolution arrays due to improved noise performance. The first APS pixel, composed of three transistors, and usually referred to as the 3T-APS pixel, marked a milestone in the evolution of image sensor architectures.

This architecture comprises a photodiode, a reset transistor, a source follower, and a selection transistor. Figure 1.16 illustrates the schematic of this basic configuration, where the selection transistor connects to a common signal, typically at the column level. The operation of the 3T-APS is as follows: first, the pixel is reset, followed by an exposure time. Finally, the selection signal activates the source follower, resulting in the establishment of a voltage on the common line, which is then converted to a digital word by a column-level *Analog-to-Digital Converter* (ADC)—which requires a calibration process to reduce FPN. Note that this working principle requires different rows to be active during distinct periods of time. The so-called rolling-shutter effect may arise when objects are moving rapidly within the image [Ohta07]. It is important to note that a notable limitation of the 3T-APS design lies in the substantial kT/C noise generated during the photodiode reset phase, rendering it less attractive for contemporary applications.

Further evolution of APS and PPD led to the development of the 4T-APS, featuring an additional transistor working as a transfer gate and a floating diffusion, fd, as depicted in Figure 1.17. This novel structure, along with the development of PPDs [Foss14], addressed several issues that CIS were facing, such as excessive dark current, lag, or kT/C noise. The transfer gate is meticulously designed to ensure complete depletion of the photodiode and the efficient transfer of all accumulated electrons to the floating diffusion region. This modification serves a dual purpose: non-destructive

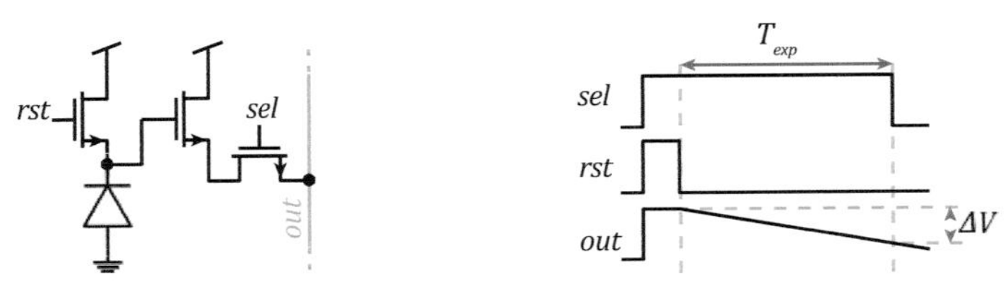

FIGURE 1.16 Architecture and Waveform Diagram of the 3T-APS.

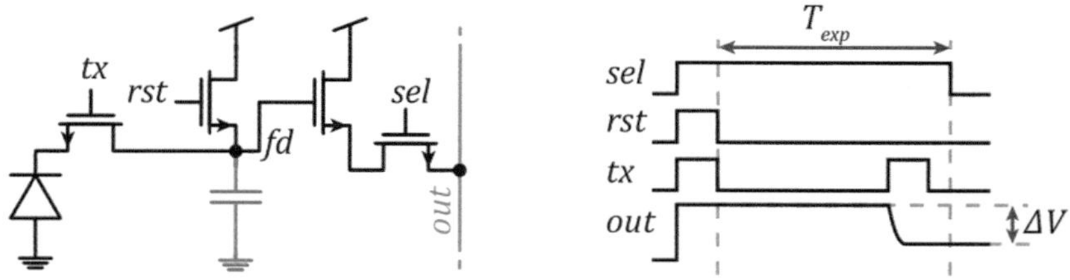

FIGURE 1.17 Architecture and Waveform Diagram of the 4T-APS.

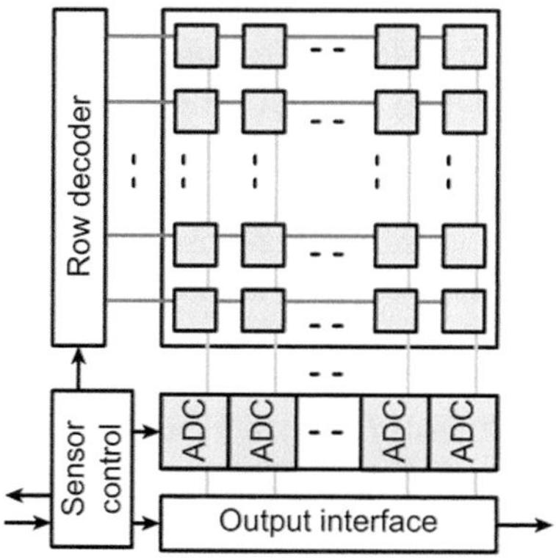

FIGURE 1.18 Basic Architecture of a simple CMOS Image Sensor with column-level ADCs.

measurement, facilitating the evaluation of the reset level post-sampling, and the potential to undertake multiple samples for improved accuracy. Also, the storage node at the pixel levels allows the implementation of a global shutter [Wang10], i.e., simultaneously capturing the signal across the entire array to mitigate the rolling-shutter effect. Nevertheless, while machine vision applications may require a global shutter to preclude motion artifacts, the majority of commercial CIS opt for a rolling shutter operation [Tele].

More complex APS architectures have been proposed to enhance their functionality through the incorporation of supplementary transistors to address challenges such as blooming [Foss14] and enable HDR capabilities [Jo14]. Moreover, alternative approaches, such as the utilization of a common-source architecture instead of a source follower, have been proposed to boost the gain of the amplification stage [Choi12; Lott11; Sato20]. Nevertheless, these strategies have not gained widespread popularity due to issues related to non-linearity and mismatches.

Having outlined the fundamental APS architectures and understanding the type of signal obtained in response to varying illumination levels, we can now delve into the basic structure of an APS-based CIS. As depicted in Figure 1.18, this architecture comprises not only a pixel array but also a row decoder, which sequentially enables the signals of individual rows. Column ADCs are usually the preferred option since they work in parallel to convert the pixel signals within the active row into a digital word. The resulting digital data is then registered, serialized, and transmitted off-chip by the output interface. It is worth noting that additional circuits are usually integrated into the sensor, such as power supply buffers and digital processing modules. In practice, nearly all sensors include pre-amplification prior to ADC conversion and mechanisms for black level correction, among other essential circuits.

Historically the single-slope ADC has been widely adopted in most CIS architectures, particularly due to the incorporation of column-level ADCs. This choice is attributed to its compact footprint and ease of calibration. However, successive-approximation-register ADCs have also found application within CIS [Funa15; Lene16; Zhan19b]. To mitigate the kT/C noise, ADCs employ a technique known as *Correlated Double Sampling* (CDS) [Kawa07], wherein both the reset voltage and signal voltage of the pixel are sampled to subtract their values. This process not only alleviates kT/C noise but also helps counteract offset mismatches and 1/f noise within the pixel. Moreover, since the measurement is non-destructive, further reduction in temporal noise is achievable through the application of correlate multiple sampling [Kawa07]. This entails sampling both the reset voltage and reset signal multiple times. Note that CDS can be executed in either the analog or digital domain, although available sensors generally opt for digital CDS implementations.

In conclusion, it is evident that APS sensors, initially oriented for imaging tasks, have established a fundamental basis firmly rooted in the domain of conventional frame-based image sensors. While they remain versatile for accomplishing numerous vision tasks, the question arises as to whether APS technology can achieve better performance than the efficiencies achievable through vision sensors integrated with focal-plane processing. Despite these considerations, APS sensors continue to be the preferred choice for industrial applications. However, APS technologies present unsolved challenges such as further reduction of noise (especially that related to the in-pixel source follower and the row temporal noise), *Light-Emitting-Diode* (LED) flicker mitigation, reduction of power consumption, to name a few. These developments are especially motivated by the integration of high-end CIS into wearable devices and automotive applications.

1.2.2 EVENT-DRIVEN VISION SENSORS

While conventional CIS architectures, as described in Section 1.2.1, focus on achieving high-quality image reproduction, event-driven sensors take a different approach by prioritizing the optimization of the overall processing efficiency [Call20a; Rodr18; Suar17]. Unlike traditional imaging, where every pixel is read out at a fixed frame rate regardless of the presence of changes in the scene, event-driven sensors only transmit information when an event occurs. This concept mimics the operation of the human eye, where our vision system is primarily sensitive to changes rather than continuously processing static information [Purv17].

One of the pioneers in the development of bio-inspired vision sensors is Carver Mead [Mead88]. His groundbreaking work in the 1980s laid the foundation for the concept of neuromorphic engineering, which involves designing systems inspired by the principles of biological nervous systems. Mead realized that traditional frame-based imaging systems were inefficient, as they often produced redundant and irrelevant data. To address this, he proposed the idea of capturing events or changes in the scene and transmitting this information directly, eliminating the need for fixed-frame rate readouts.

The work of Carver Mead led to the development of sensors that only respond to temporal [Lich08] or spatial contrast [Boah91]. This event-driven approach drastically reduces the amount of data generated by the sensor, leading to several advantages. Firstly, it reduces the computational load required for subsequent processing stages, as only relevant data is transmitted. Secondly, the lower data rate allows for reduced power consumption and more efficient use of resources. Thirdly, event-driven sensors have a remarkably low response latency, making them well-suited for tasks that require rapid reaction to changes, such as robotics and autonomous vehicles.

Figure 1.19 illustrates the temporal evolution of output from frame-based and event-based sensors. The output of a frame-based sensor operates synchronously and is bound by the frame rate, whereas the event-based sensor communicates events asynchronously. An event is related to an occurrence in the scene, such as a sudden change in pixel intensity or a change of pixel values within a local neighborhood. Figure 1.19 (c) showcases the representation of the output signal of an individual pixel, which can be viewed as a sequence of events, each with a corresponding timestamp.

Advantages of time-encoded signals

The time-encoded paradigm employed in event-driven sensors offers unique advantages pertaining to the nature of the signals themselves, particularly in terms of achieving higher dynamic range and improved signal representation. In this paradigm, events are represented as individual time stamps when changes occur, rather than as part of a continuous stream of image frames. These events can be seen as Dirac pulses distributed over time.

One of the key benefits of time-encoded signals is their potential for achieving almost infinite dynamic range. Unlike traditional frame-based sensors that are constrained by the resolution of their ADCs, time-encoded signals do not present this problem. When a significant event occurs, it is encoded with a precise time stamp, allowing even extremely large intensity variations to be accurately captured. This is because the time-encoding paradigm can effectively represent events that span

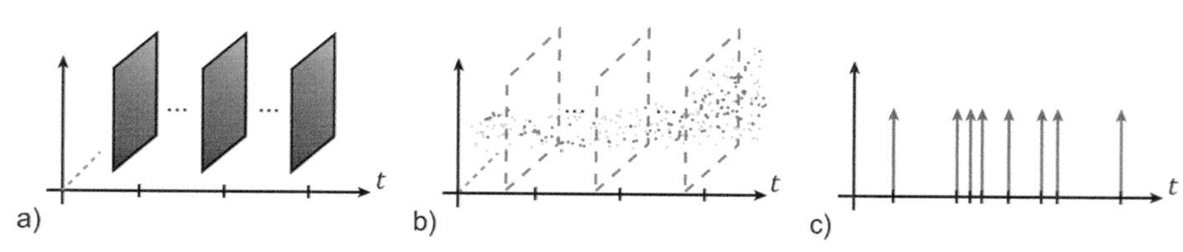

FIGURE 1.19 Temporal Evolution of output from: a) Frame-Based Sensor. b) Event-Driven Sensor. c) a single pixel from b), represented as a sequence of timestamps indicating occurrence of events.

multiple orders of magnitude in intensity, ensuring that even the smallest and largest changes are captured with high fidelity [Lene17].

Time-encoded signals also offer a distinctive advantage due to their inherent representation of discrete events. Unlike conventional image signals where the entire frame is associated with the same time interval, time-encoded signals of each event may be recorded as a distinct time stamp for each event, which is transmitted asynchronously [Boah00; Culu03; Lich08]. This feature allows for an exceptionally high temporal resolution, facilitating the precise capture and representation of dynamic changes with a low latency. Consequently, time-encoded signals excel in capturing transient events and rapid temporal variations with enhanced fidelity, making them particularly advantageous in applications requiring high-speed or dynamic signal acquisition and analysis, since the acquisition of information is not constrained by a frame rate.

The benefits of this property extend to scenarios where rapidly changing intensities or high-frequency variations are crucial. For instance, in scenes with extreme lighting changes or sudden motion, time-encoded signals can accurately represent these dynamics without the need for high frame rates or complex exposure control mechanisms. This makes event-driven sensors particularly suitable for applications requiring robustness in challenging lighting conditions and responsiveness to rapid scene changes, further emphasizing their advantages in specialized vision tasks.

Temporal contrast sensors: The principles of Dynamic Vision

Temporal Contrast Sensors, often referred to as dynamic vision sensors (DVSs), introduce a distinctive approach to capturing visual information. Originally envisioned by Jörg Kramer [Kram02a; Kram02b] these sensors were designed to transmit events to an external receiver in reaction to fluctuations in illumination levels, whether positive or negative, at individual pixels. Unfortunately, Jörg Kramer passed away, but Lichtsteiner and Delbruck continued his work, ultimately leading to the development of what we now recognize as a DVS sensor [Lich08].

Figure 1.20 (a) illustrates the fundamental implementation of a DVS pixel. This architecture consists of several crucial components that collectively enable operation of the pixel. The photodiode is connected in a configuration where the output

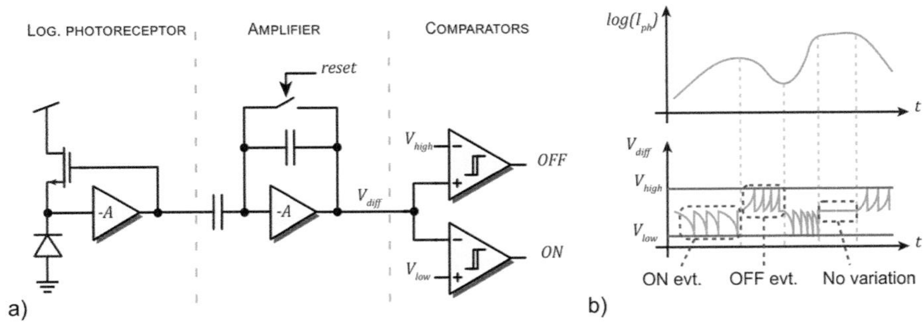

FIGURE 1.20 a) Schematic of the original DVS pixel. b) Temporal evolution of the signals.

voltage of the photoreceptor, V_{ph}, evolves logarithmically with the photocurrent. This aspect is crucial for the implementation of a DVS, as it ensures that the magnitude of changes is relative to the current operating point. In contrast, a linear voltage response would be highly sensitive to high illumination but considerably less sensitive to low illumination variations. Then, a capacitive amplifier is responsible for detecting and amplifying variations in V_{ph}. To achieve this, the amplifier is initially reset, setting its output equal to the quiescent voltage of the amplifier (or a defined voltage in the case of a differential amplifier) which is deliberately chosen between the thresholds of the comparators, V_{high} and V_{low}. Subsequently, any change in V_{ph} will be magnified and result in a corresponding variation in V_{diff}. If this amplified signal crosses the threshold of either comparator, an event will be triggered. It is important to note that the handshake circuitry necessary for transmitting this event outside of a pixel array has not been included in this description. Figure 1.20 (b) illustrates this process.

The output of the DVS pixel is a continuous stream of events, with each event containing information about the timing of the event, and whether it was a positive or negative change. This event-driven output stands in contrast to conventional frame-based sensors, which produce entire frames of data irrespective of the presence or absence of significant changes in the scene.

The temporal contrast sensing paradigm of DVS pixels confers several advantages. Since the sensors respond to relative changes rather than absolute light levels, they offer a significantly higher dynamic range, enabling them to capture a broader range of luminance variations within a scene. Additionally, these sensors drastically reduce the data volume and computational burden associated with traditional frame-based sensors.

This approach has found applications in areas such as robotics, motion detection, and object tracking, where real-time responsiveness and efficient data representation are essential. Also, DVS architecture can be combined with conventional APS to mitigate blur artifacts [Guo23a; Koda23].

Spatial contrast sensors

Edge detection constitutes a fundamental process in the field of image processing. This operation involves comparing the intensity value of a pixel with that of its neighbors. While various contrast definitions have been proposed, such as the Michelson or Weber contrast [Peli90], they all share a common requirement: obtaining the values of all neighboring pixels prior to computation.

Traditionally, the contrast operation is achieved through convolution utilizing different kernels in both horizontal and vertical directions. However, this operation involves a significant computational overhead, thus emphasizing the importance of designing spatial contrast detectors that can execute this processing directly within the focal plane [Gott09; Rosk93]. This not only enhances the latency of the final application but also alleviates the computational burden.

A conceptual implementation, adapted from [Mead88], is illustrated in Figure 1.21. In this example, the intensity values of pixels are encoded within the voltage domain. Subsequently, the average intensity of the surrounding neighborhood is computed through the utilization of a resistive grid. By comparing the average intensity value

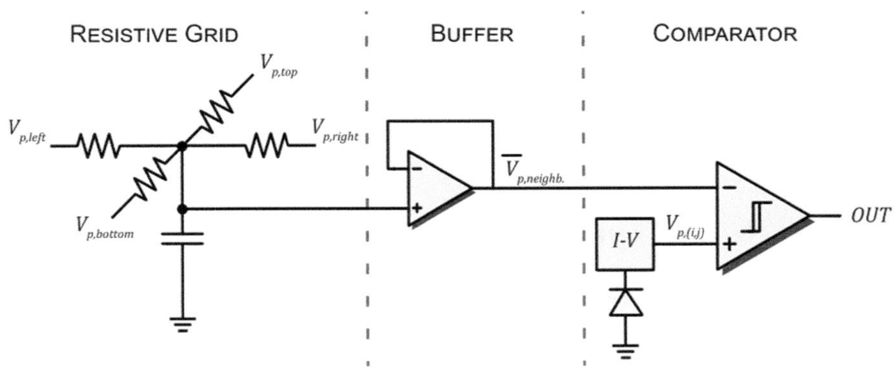

FIGURE 1.21 Conceptual implementation of a spatial contrast detector.

of the neighborhood with the absolute value of the intensity of the pixel in question, edges can be detected and even quantized, provided the difference is measurable.

Numerous authors have proposed diverse architectures in this domain [Genn23; Park23; Soel16]. However, none have yet attained the level of maturity and versatility exhibited by conventional image sensors in real-world applications. Consequently, spatial contrast detection remains a challenge that demands resolution within the field of image processing.

Emergence of Octopus Retinas: Beyond Frame-Based Imaging

Spiking luminance sensors, also known as octopus retinas [Culu03], are event-based sensors proposed to extract the absolute luminance level, as traditional CIS do. However, these sensors depart from the concept of fixed frames and instead output a continuous pulse train for each pixel, encoding the luminance level in the frequency domain. This novel approach draws inspiration from biological systems, processing visual information in a manner analogous to neurons. In the octopus retina-inspired architecture, each pixel functions as an independent unit that generates a pulse in response to changes in illumination. These pulses, analogous to neural spikes, are asynchronously transmitted, allowing the sensor to convey information more naturally, similar to how the human visual system works.

The octopus retina represents a departure from the traditional frame-based paradigm, enabling continuous data streaming in response to dynamic scenes. This approach offers a unique way to capture illumination and transmit information efficiently, avoiding the redundancies associated with frame-based imaging.

Figure 1.22 (a) illustrates the most straightforward implementation of an octopus retina pixel, while Figure 1.22 (b) depicts the temporal evolution of the operational signals. The process works as follows: the photodiode is initially reset to the pixel reference voltage, V_{pix}, starting the charge collection. When the voltage of the photodiode, V_{ph}, reaches the comparator threshold, V_{th}, the comparator toggles, generating a voltage spike and initiating a self-reset action. Thus, the luminance level is encoded in the frequency domain, represented by the expression:

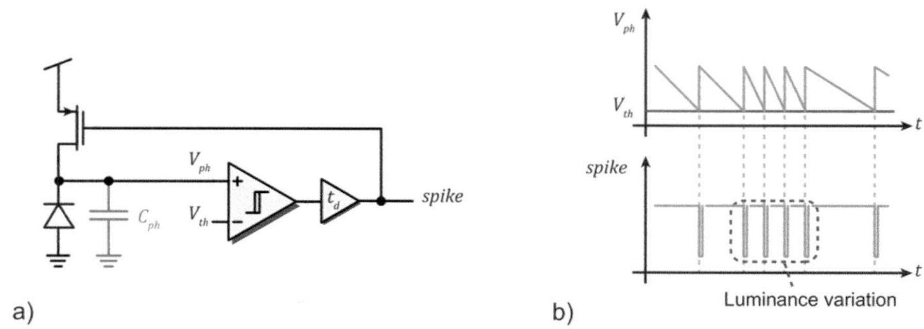

FIGURE 1.22 a) Pixel schematics of an octopus retina. b) Temporal evolution of the signals.

$$J_{spike} = \frac{I_{ph}}{C_{ph}\left(V_{pix} - V_{th}\right)} \tag{1.10}$$

One notable advantage of the octopus retina lies in its ability to achieve a wide dynamic range through frequency-domain encoding. While frame-based CISs implement techniques such as combining subframes with differing exposure times or in-pixel tone-mapping [Varg15] for HDR imaging, octopus retina inherently present a HDR operation. We can consider that the octopus retina dynamically adjusts the exposure time of each pixel based on the intensity level. This adaptation arises from the significance of an event in this context. In the octopus retina, an event means that the V_{ph} reaches V_{th}. As a result, the exposure time is longer for dark pixels than for bright pixels. Indeed, an event is equivalent to the collection of a specified number of electrons, N_e, as given by the equation:

$$N_e = \frac{C_{ph}\left(V_{pix} - V_{th}\right)}{a} \tag{1.11}$$

This approach is also interesting in terms of noise. By setting N_e sufficiently large, photon shot noise is dominant, thus making the readout noise irrelevant. Additionally, the noise contribution is further reduced when multiple spikes are employed to process information. However, this asymmetric exposure presents certain drawbacks. Firstly, it can result in intensity-dependent motion artifacts in dark areas, where exposure times are excessively long.

In conclusion, octopus retina sensors offer a unique approach for capturing and encoding luminance information through temporal spikes, enabling efficient bandwidth utilization. These sensors excel in scenarios where dynamic adaptation of exposure times is beneficial, yielding remarkable results in terms of efficiency [Culu03; Gome23a]. However, their current limitations lie in their compatibility with traditional frame-based applications commonly found in industrial tasks. The reconstruction of frames from spiking data may not yet match the efficiency of direct frame capture by CIS. Nevertheless, as research advances in the development of neuromorphic processors that directly utilize spiking data for decision-making, the potential of the octopus retina becomes more promising. In this evolving landscape, octopus retinas could emerge as a powerful tool, ready to offer exciting and innovative solutions.

Event-driven readout

Event-driven architectures, as discussed earlier, feature specialized pixels designed to generate events associated with different meanings, e.g., temporal contrast, spatial contrast, or intensity encoding. Notably, these architectures operate independently from the conventional frame-based paradigm. Consequently, pixels operate autonomously, generating information in an asynchronous manner, rather than orchestrated by sequential scanning either during acquisition or during data downloading.

The *Address Event Representation* (AER) protocol is a widely adopted technique for transmitting these events [Boah00; Gome23a; Lich08]. Figure 1.23 shows the transmission and reception of data using the AER protocol. In this protocol, an AER link is utilized to communicate the address of each pixel. This allows the receiver to reconstruct the original signal by referencing the associated timestamp for each reception. The req signal denotes the presence of valid data in the address bus, while the ack signal confirms the successful storage of the address.

Figure 1.24 (a) illustrates an example of a sensor implementing the AER protocol, adapted from [Boah00]. Alongside the pixel unit, three critical components are integral to this implementation. Firstly, the AER communication logic takes charge of the handshaking. It manages the necessary signals for communication between the pixel units and the arbiter tree. The arbiter tree determines which unit accesses the readout channel, typically prioritizing the element that initiated the access request first. Subsequently, upon granting access to an element, the arbiter tree enables a signal confirming this access to the address encoder, which records the address of the element. In 2D arrays, this circuitry is replicated in both dimensions, performing the arbitration process consecutively in each. Figure 1.24 (b) depicts the waveform diagram of this architecture, where req_x and req_y are the request signals in each

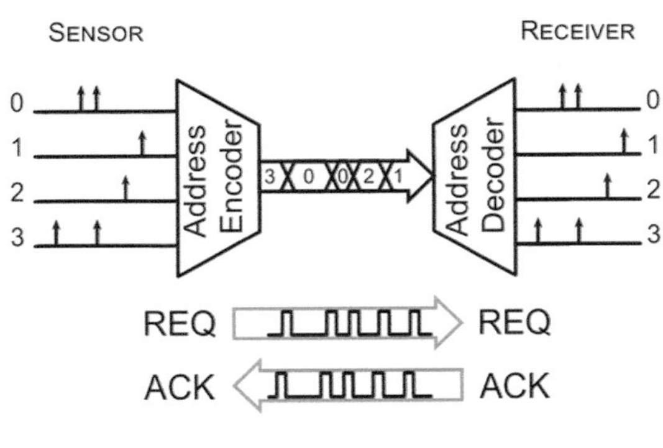

FIGURE 1.23 Transmission and reception of information using the AER protocol.

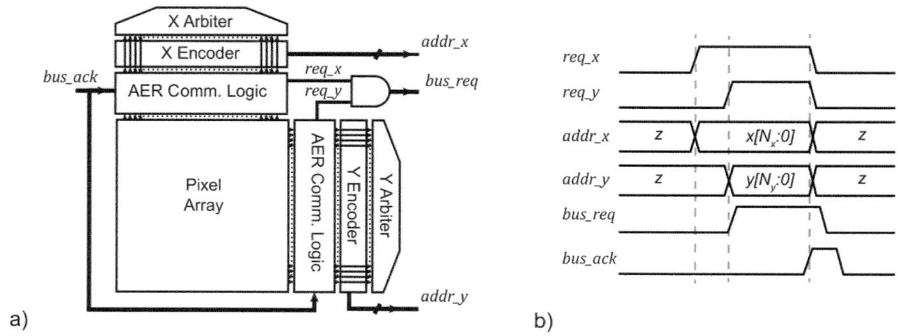

FIGURE 1.24 a) Sample architecture of sensor implementing the AER protocol b) Waveform diagram.

dimension. Therefore, the global request signal *bus_req* is implemented by the AND operation of both signals.

However, this implementation introduces a significant drawback. Information is conveyed through the timestamp of event transmissions, making the system sensitive to delays in event transmission, congestion on the readout channel, and dead times in the external receiver responsible for timing. These factors manifest as errors in measurements. While this effect may resemble random noise in some cases, it represents a scene-dependent signal distortion. This non-deterministic issue becomes particularly pronounced as array spatial resolution increases, leading to higher loads on common lines and more concurrent pixel requests.

To address this challenge, image sensor designers have proposed various techniques. These solutions range from pixel grouping and clustering [Son17], to reverting to traditional scan-based readout methods [Suh20]. This may involve scanning only rows where events have been generated [Guo23a] or employing high-speed scanning with different time intervals for distinct event conditions [Koda23].

1.2.3 Time-of-Flight sensors

In this section, we delve into the circuit implementations of ToF techniques discussed in Section 1.1.3. ToF sensors employ the principle of measuring the time taken for a light signal to travel to an object and back. By exploring the underlying electronic architectures and signal processing methods, a proper understanding of the diverse approaches employed in ToF sensor designs is provided.

Indirect time-of-flight pixels

As discussed in Section 1.1.3, iToF techniques rely on the computation of phase differences between emitted and reflected light. This is achieved by enabling distinct time intervals for charge accumulation. Subsequently, iToF is derived by subtracting and dividing the corresponding charges.

To implement this functionality, iToF pixels incorporate multiple storage nodes, each comprising a floating diffusion and a transfer gate. Figure 1.25 depicts a 2-tap

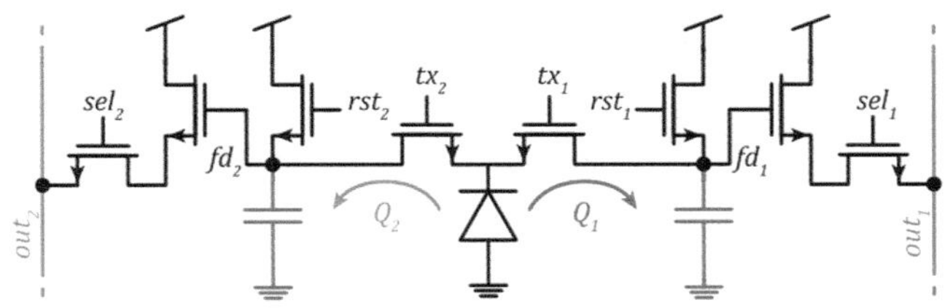

FIGURE 1.25 Schematics of a 2-tap pixel for iToF.

pixel. This architecture can store charge corresponding to two different phases, thus acquiring the entire frame after two sub-frames. Extending the number of taps allows detecting the signal corresponding to different phases with fewer exposures.

While it is evident that an increase in the number of taps improves acquisition speed, augmenting the taps can be challenging due to factors such as geometric constraints and pixel pitch limitations. Additionally, there is a trade-off, as it leads to a reduced fill-factor attributed to the incorporation of extra circuitry and routing. However, this aids in mitigating motion artifacts. In practice, the number of taps is typically limited to 2 or 4, with a maximum reported value of 8 [Shir20].

Alternatively, iToF can be implemented utilizing SPADs by counting photons in each phase. A direct implementation might employ a SPAD integrated with a counter activated exclusively during the corresponding phase. However, this simplified approach is susceptible to motion artifacts and exhibits lower power efficiency due to increased laser exposures.

To address this challenge, the concept of tap can be again of interest. A recent work has proposed multi-tap SPAD-based pixels, incorporating multiple counters in the analog domain to accumulate photon counts across two phases per exposure [Park21]. Additionally, if the accumulation process is executed digitally, phase subtraction can be computed at the pixel level by integrating up-down counters.

Direct time-of-flight pixels

While SPADs allow implementing iToF, their primary advantage lies in their capability to execute dToF measurements. However, despite the apparently straightforward nature of time measurement, factors like resolution requirements, photon statistics, and background illumination introduce complexity into the problem.

Pixels implementing dToF pixels can be classified into two main groups. The first group includes pixels designed to trigger a TDC [Vorn17], as depicted in Figure 1.26 (a). Note that TDCs may be shared by columns or clusters to reduce power and area consumption, at the expense of possible collisions and a more complex arbitration scheme. The second group comprises gated pixels that are enabled for a brief period, temporally shifted with respect to the laser pulse. Figure 1.26 (b) represents a conceptual implementation, where a memory element is included in the pixel to indicate whether a photon was detected or not. Note that the simplest implementation of this element may be a floating diffusion [Mori20a], but it can also be a counter that directly computes the bin data [Gyon23; Rocc20]. It is evident that gated pixels

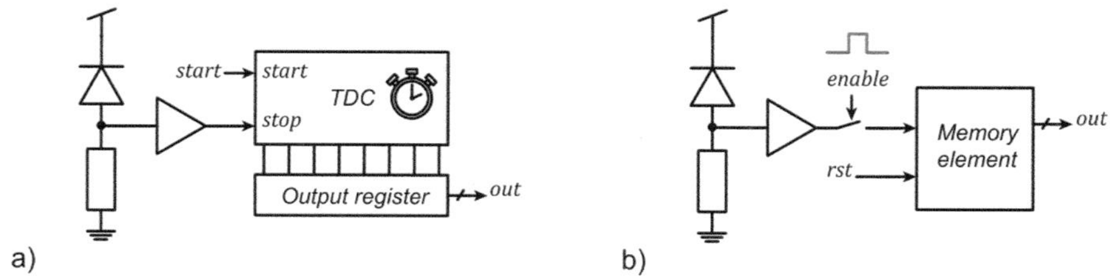

FIGURE 1.26 a) dToF pixel including per-pixel TDC. b) time-gated dToF pixel.

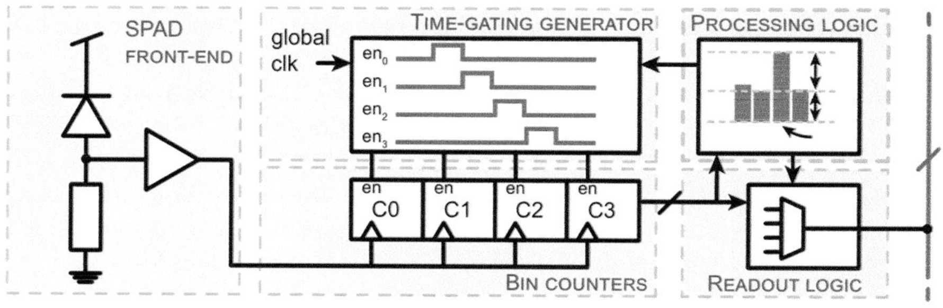

FIGURE 1.27 Time-gated dToF pixel including advanced circuitry.

offer a smaller pixel footprint, albeit at the expense of sacrificing the high-resolution capabilities offered by per-pixel TDCs. Additionally, gated pixels require extra laser exposures, resulting in increased latency and power consumption, but they offer a versatile architecture that benefits from the increasingly integration capacity of more advanced technological nodes.

In both approaches, the construction of histograms [Vorn19] is imperative to account for photon statistics. However, computing histograms outside the sensor leads to a vast volume of data to be transferred, stored, and processed. Hence, additional circuitry is commonly integrated at the pixel level to build histograms and overcome these limitations. Notably, in recent years, gated pixels have emerged as the preferred solution, mainly because they demonstrate superior scalability in terms of power consumption when increasing the number of pixels.

Figure 1.27 depicts the different blocks that a gated dToF pixel may include. In addition to the single-photon detector, the pixel may feature a time-gating generator. This component generates the gate or gates of the pixel to attribute photon detection to specific bins. Moreover, the pixel may incorporate multiple counters to facilitate bin accumulation, along with processing circuitry responsible for tasks like peak detection, control of the time-gating generator, to name a few [Gyon23; Rocc20]. Consequently, the output of a dToF pixel may comprise raw bin data, peak bin information, or even the CoM of the full histogram.

Achieving a wide range of distances with fine resolution requires numerous laser exposures. This translates into a large number of bins in the histogram, which is impractical to implement at the pixel level. In reality, the information is only related

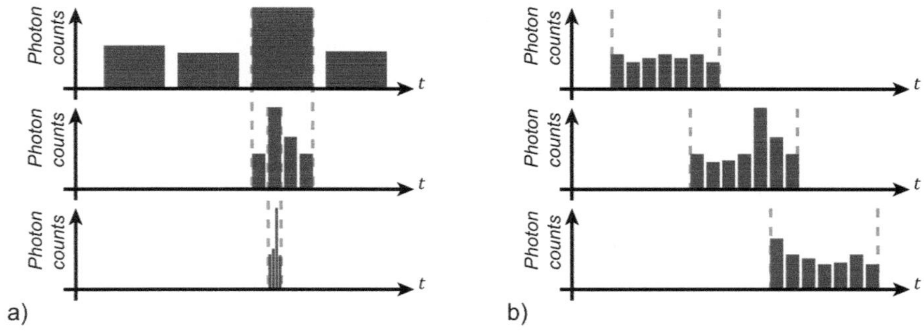

FIGURE 1.28 Partial histogramming techniques: a) zooming, and b) sliding.

to a limited subset of histogram bins, rendering a full histogram sweep inefficient. To address this challenge, two main techniques are employed in the literature: zooming and sliding [Gyon22].

Histogram zooming involves initially performing a coarse histogram to extract the peak. Subsequently, the time-gating generator is individually adjusted to enhance resolution in the region corresponding to the peak of each pixel, as demonstrated in Figure 1.28 (a), iteratively sweeping different positions in each pixel. This process continues until the desired resolution is achieved. Conversely, sliding techniques entail continuously shifting the timing interval until a peak is detected, as shown in Figure 1.28 (b). This yields the *Most Significant Bits* (MSBs) corresponding to the interval position of each pixel, while the *Least Significant Bits* (LSBs) are computed using the Center of Mass algorithm. Recent studies have demonstrated that both zooming and sliding techniques hold merit, with the former excelling in low-background illumination scenarios and the latter proving advantageous in the opposite scenario [Tane22].

1.3 SPECIFICATION AND CHALLENGES FOR LIDAR SYSTEMS

This section is dedicated to the rigorous examination of the specifications and challenges pertinent to LiDAR systems. LiDAR technology, utilizing laser pulses for precise distance measurement and detailed 3D mapping, plays a pivotal role in numerous applications, including autonomous navigation and environmental analysis [Yole23b]. By delving into the specific technical requirements and difficulties inherent to LiDAR systems, this section provides an understanding of the intricate considerations that govern the development and implementation of this critical sensing technology.

1.3.1 LiDAR APPLICATIONS

LiDAR technology has found its place in a diverse range of applications where precise depth sensing is crucial. These applications span AR, face recognition, autonomous driving, robotics, and surveillance, among others.

One of the most prominent applications of LiDAR technology is in autonomous driving. LiDAR systems map the surrounding environment with precision, extracting crucial information for understanding the scene. This depth perception is pivotal for detecting obstacles, pedestrians, road signs, and other vehicles. Notably, LiDAR is fundamental not only for autonomous driving but also for *Advanced Driver Assistance Systems* (ADAS). ADAS enhances the driving experience by aiding the driver in crucial decision-making processes, such as applying brakes when necessary to prevent accidents.

In applications like face recognition and AR, LiDAR technology is leveraged to enhance user experiences when interacting with virtual elements [Gene22; Kim23; Xie23]. Face recognition enables secure authentication by identifying unique facial features, using them as a key for various applications. AR, on the other hand, aims to enrich everyday functionalities. Companies are increasingly investing in the development of smart glasses equipped with multiple cameras to map the environment. Interactive displays on the lenses provide users with valuable information.

It is important to note that different applications have distinct requirements. For instance, face recognition demands fine resolution on the order of millimeters to discern details that differentiate individuals, allowing for measurement times in the hundreds of milliseconds or even seconds. Conversely, applications like autonomous driving require low latency to swiftly respond to unforeseen dangers. Table 1.2 provides a summary of the key requirements for various LiDAR applications, including operational range, resolution, processing speed, and power consumption.

TABLE 1.2
Requirements for different LiDAR applications.

Application	Description	Range	Resolution	Speed	Power budget
Augmented Reality	Enhancement of virtual object's interaction with real world.	Low	High	Low	Low
Autonomous driving	Mapping and perception of surroundings.	Medium-High	Medium	Medium	Medium
Environment mapping/3D scanning	Creation of 3D detailed maps of environments.	High	Low	Low	High
Surveillance	Accurate and real-time detection of objects and individuals.	Medium	Low-Medium	Low	Medium
Industrial automation	Quality control and automation of processes.	Low-Medium	Medium-High	Medium-High	High

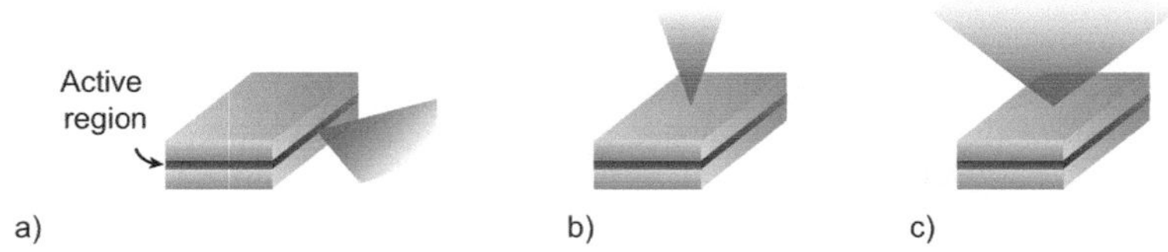

FIGURE 1.29 Different emitter technologies: a) edge-emitting laser, b) vertical-cavity surface-emitting laser, and c) light-emitting diode.

1.3.2 Emitter Technologies

While the significance of the receiver in ToF system is undeniable, it is imperative to recognize that the role of the emitter in a LiDAR system is, at least, as important as the receiver. The technology and characteristics of the emitter present a substantial influence over the performance of the entire system. Laser parameters such as maximum power, delay, and dynamics significantly impact the measurement process. For instance, a low-power laser restricts the operational range to a few meters as only a limited number of photons can return when targets are far away. Additionally, the rise and fall times dictate the laser's pulse width, an attribute that ideally should be minimized.

Among the diverse emitter technologies, several stand out as commonly utilized in LiDAR systems. These technologies include *Vertical-Cavity Surface-Emitting Lasers* (VCSELs), *Edge-Emitting Lasers* (EEL), and LEDs [Main13], each possessing distinct advantages and limitations. Figure 1.29 depicts these technologies, illustrating beam characteristics.

VCSELs offer notable advantages in terms of beam quality, allowing for precise focusing of the laser beam. Moreover, they are recognized for their relatively low testing costs and ease of fabrication. However, it is important to mention that VCSELs may present limitations concerning maximum power output, potentially affecting their operational range, particularly in long-range LiDAR applications. Also, power efficiency is usually lower than alternative technologies.

EELs are characterized for emitting light from the edge of the substrate. They possess the capability to generate high peak powers, a characteristic advantageous in scenarios where intense laser pulses are a requisite. However, they may exhibit challenges concerning beam divergence, requiring the inclusion of additional optical components to achieve the desired beam characteristics. Also, they are associated with higher power consumption and may require sophisticated cooling systems to manage heat dissipation effectively.

Lastly, LEDs are a cost-effective option commonly utilized in low-power LiDAR systems. They provide advantages such as low power consumption and availability in various wavelengths. LEDs are relatively easy to drive and have longer lifetimes compared to laser-based emitters. They may be employed in short-range LiDAR applications, where precise distance measurements are not the primary requirement. However, LEDs generally have lower power output and broader beam divergence, limiting their suitability for long-range and high-precision LiDAR applications that demand narrower beam profiles and higher accuracy.

Another critical characteristic influencing the design of LiDAR systems is the level of integration of the laser. While certain applications, such as industrial automation, may accommodate larger and more substantial laser devices, others, like augmented reality implemented in smart glasses, demand integrated solutions. In such cases, there is a preference for emitters and receptors to be seamlessly incorporated onto the same board. Notably, among the commonly used emitter technologies, VCSELs and laser diodes offer superior potential for integration due to their compact size and ease of fabrication. EEL may pose more significant challenges in achieving high levels of integration, potentially requiring additional engineering efforts to meet the stringent spatial constraints of compact applications.

1.3.3 Size, Weight, and Power

In the context of LiDAR systems, *Size, Weight, and Power* (SWaP) optimization may be a critical metric, especially in applications like augmented reality or those reliant on battery-powered systems. The components within a LiDAR system can impose constraints on these metrics. For instance, scanning LiDAR systems often involve bulky components and moving parts, making integration into compact devices challenging. This is why Flash LiDAR is favored in space-constrained applications.

When mechanical components are eliminated, lasers become the primary source of power consumption. As discussed earlier, emitters must strike a balance between factors like footprint, power efficiency, and beam quality. The integration of appropriate driver circuits is crucial. These drivers must exhibit both high power efficiency and the capacity to deliver the substantial current required by laser emitters, frequently within short pulse widths. This translates to a notable energy flowing through the devices, inevitably resulting in thermal generation. Hence, there is sometimes a need to incorporate additional components, such as heatsinks or stabilization circuits, which can have repercussions on power efficiency and area consumption.

However, the challenge of SWaP optimization extends beyond the emitter itself. Ideally, the receiver should be able to compute ToF by using a single laser exposure. However, in practice, many laser exposures are required to obtain a measurement. As explained in Section 1.2.3, circuit techniques can be improved to reduce the number of exposures for a given range. However, an always-on ToF can constitute a substantial power consumption. An efficient solution that some researchers have evaluated consists of including motion-detection mechanisms to efficiently trigger the ToF mode of the sensor only when variations on the scene are detected. Although DVS technologies are becoming mature in the field of conventional image sensors, it is still a challenge when it comes to performing this operation when using SPADs as sensing devices.

Moreover, additional components are essential in these systems. For instance, SPAD devices operate at high voltages, requiring the inclusion of DC-DC conversion and regulation circuitry in many cases. This addition increases the overall power consumption and contributes to the footprint of the entire system. Overcoming this

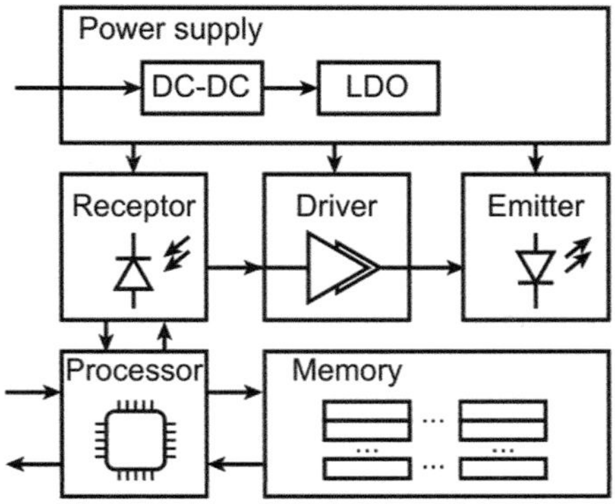

FIGURE 1.30 Example of LiDAR system including receptor, emitter, power supply management, processor, and memory.

challenge involves exploring on-chip methods for generating these high voltages, potentially employing calibration or feedback schemes to account for *Process, Voltage, and Temperature* (PVT) variations.

Additionally, the vast volume of data generated by SPADs poses a significant consideration. This necessitates robust processing and storage resources within the system, which can, in turn, impact power efficiency and system size. Addressing this challenge involves the incorporation of circuitry that pre-processes the data, reducing the overall data throughput and consequently mitigating the impact on SWaP metrics.

Figure 1.30 provides an overview of the various components that a SPAD-based Flash LiDAR system may incorporate for autonomous operation.

1.4 OVERVIEW OF THE STATE OF THE ART

Up to this point, we have delved into fundamental architectures for SPAD-based dToF detectors, all of which rely on the integration of TDCs. These TDCs can take the form of circuits that directly yield a digital value corresponding to the ToF or they can adopt a time-gated configuration, allowing for the definition of different temporal bins. The latter option aligns well with the concept of histogramming, as these gates can separate photon detections into various bin counters. Additionally, we have discussed in Section 1.2.3 that various techniques have been proposed to diminish the number of laser exposures needed for a measurement, involving in-pixel circuitry to compute partial histograms during operation.

However, modern applications incorporating smart sensors in battery-powered devices are demanding always-on operation of such sensors to continuously gather information from the environment. Consequently, there is a growing interest in implementing intelligent SPAD-based sensors able to extract pertinent features from the scene, optimizing both information processing and power consumption. Specifically, the implementation of a motion-triggered ToF system has emerged as a focal point in the pursuit of these objectives.

This section showcases SPAD-based sensors implementing vision tasks, presenting their key principles and results. Furthermore, this section provides a review of the evolution of event-driven readout architectures in SPAD-based sensors. Note that the results and descriptions provided for each work have been simplified for the sake of clarity. In most cases, the actual circuits and functionalities are more complex than what is described here. For a more detailed understanding of the systems and circuits, refer to the original work.

1.4.1 SPAD-based Vision Sensors

Vision sensors based on linear-mode photodiodes have been reported both in academic research and in the form of industrial products [Tele18]. However, the development of vision sensors based on SPAD devices is an area that still requires significant exploration. The digital nature of their signals and the use of advanced technology nodes are two key elements that open the opportunity for enabling the implementation of advanced tasks that were not possible in the past.

Although vision tasks may include edge detection, segmentation, or classification, to name a few, the only vision task that has been tried to implement in the context of SPAD-based sensors is the implementation of motion detection. Conveying the concept of dynamic vision to the field of SPAD-based sensors may be a solution to reduce the data throughput of such devices. However, the digital and random nature of photon detection poses specific challenges for their implementation.

In particular, two sensor architectures proposed by researchers of the University of Edinburgh are of interest [Gyon23; Rocc20]. These architectures are based on the storage of information within the pixel to implement frame-comparison techniques and eventually detect motion. The first one [Rocc20] stores information related to the photon count of the last frame, while the one reported in [Gyon23] stores the information about the depth. Both sensors excel in terms of data throughput reduction and provide an efficient way to detect motion using SPAD-devices, mitigating the problem of the vast amount of data that they generate. Both architectures will be described in the subsequent sections.

Photon-count-based motion detection

The sensor proposed in the work in [Rocc20] introduces a distinctive vision mode, wherein the output of the sensor constitutes a frame of 2-bit values. In this configuration, each pixel's value can take one of three possible states: either no motion is detected, a positive contrast is observed, or a negative contrast is detected.

Figure 1.31 depicts a simplified schematics of the pixel described in [Rocc20]. To facilitate the frame comparison operation, each pixel incorporates two 15-bit counters. Consequently, one counter is active during the current exposure period, while the second serves as a memory element, retaining the photon count from the previous frame. Finally, the pixel array is sequentially scanned, and the data from both counters is relayed to the column vision processor.

The column vision processor assumes the responsibility of processing the data to ascertain if there has been an intensity variation since the last frame. This is achieved

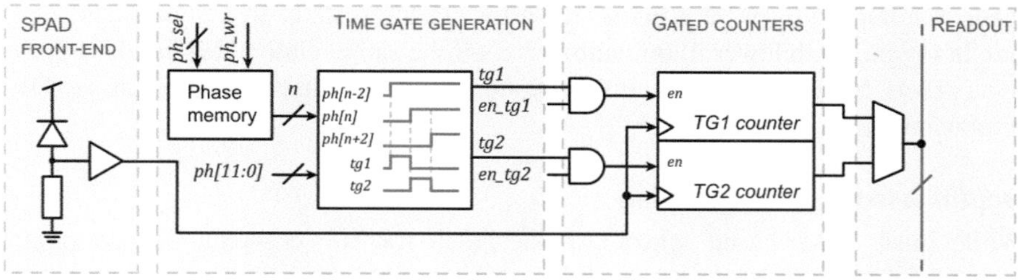

FIGURE 1.31 Simplified pixel schematics of the sensor reported in [Rocc20].

by subtracting the values from both counters and subsequently comparing the result to a predetermined threshold. It is crucial to note that this threshold must be contingent upon the illumination level to account for relative variations rather than absolute ones. Thus, the output of the column processor can be expressed as:

$$out = \begin{cases} 0 & if \ -\frac{1}{2^n} < c < \frac{1}{2^n} \\ 1 & if \ c \geq \frac{1}{2^n} \\ 2 & c \leq -\frac{1}{2^n} \end{cases} \tag{1.12}$$

where c is the temporal contrast computed by means of the counts of the former frame, $count_{F1}$, and the latter frame, $count_{F2}$. Thus, c is expressed as:

$$c = \frac{count_{F2} - count_{F1}}{count_{F1}} \tag{1.13}$$

One of the noteworthy advantages of this sensor lies in its capability to operate in a low-power always-on mode, yielding data pertinent to motion detection. This data is subsequently filtered in the subsequent processing stage, i.e., in a *Field Programmable Gate Array (FPGA)*, and if motion is detected, it triggers the ToF mode to update the depth value. Pixel counters are repurposed in ToF mode to generate 2 bins per frame, building a 6-bin histogram after reading out 3 frames. These bins are determined by the time gate generator, which generates the time gates from 12 clock phases that are globally distributed.

However, this approach does present a limitation inherent to the acquisition process. Since the system operates synchronously and adheres to a defined exposure time, the motion detection capability is constrained by the frame rate of the system. In the specific implementation of [Rocc20], the maximum frame rate reported is 500 fps, resulting in a latency of at least 2 ms. This latency remains constant regardless of the number of photons collected by the pixel. Additionally, this approach introduces another challenge: an illumination-dependent photon count in the comparison stage leads to a non-uniform SNR across the array. This translates to a higher false event rate in regions with lower illumination levels if the same relative threshold is applied to all pixels. Nevertheless, this noise behavior aligns with that of existing analog DVS implementations.

Depth-based motion detection

While most SPAD-based sensors are designed for ToF computation and photon counting operations, an alternative approach is possible. Instead of comparing photon counts from one frame to the next to detect changes in intensity, it is feasible to detect variations in depth for motion detection within the scene.

This strategy aligns well with the sliding approach for histogram construction since histogram peaks can be tracked to detect when the peak bin varies. This strategy was reported in [Gyon23]. Essentially, during each exposure, the pixel accumulates information corresponding to eight bins of the histogram. Subsequently, the sweep range is shifted, and the process is repeated until a peak is detected. The detection of a peak is also executed at the pixel level, considering Poisson statistics. The output of the sensor is then chosen to be either the data of the histogram peak, complete information on all bins, or the CoM computed in the column processor.

While the sensor is not originally designed to operate as a motion detector, it implements a smart readout mode where only pixels detecting a moving peak are read out. This readout mode, combined with the on-chip CoM computation, offers a clear advantage in terms of data throughput and power consumption, as irrelevant data is not transmitted off chip.

Nevertheless, depth-based motion detection inherently requires depth computation. Consequently, redundant laser exposures may be conducted, making it impossible to implement a motion-triggered ToF mode. Additionally, the synchronous nature of the sensor imposes limits on achievable temporal resolution.

1.4.2 Event-driven readout in SPAD-based sensors

The inherent asynchronous nature of photon reception and therefore, of SPAD detectors, have motivated many researchers to design and implement alternative SPAD-based architectures that benefit from the advantages of event-driven readout schemes. Indeed, the work in [Afsh19] demonstrated the advantages of event-based processing of SPAD sensors by using a dataset from a frame-based sensors. Researchers have proposed asynchronous approaches for SPAD-based sensors [Berk15; Sham18]. However, the experimental results for these latter have not yet been reported. Nevertheless, the implementation of a true asynchronous SPAD-based image sensors scalable and suitable for imaging applications that require a moderate resolution have not been reported yet.

This section delves into event-driven architectures utilizing SPADs as sensing devices. It outlines their strengths, highlighting why they excel in their respective applications, while also shedding light on the challenges faced when attempting to scale them for general vision tasks.

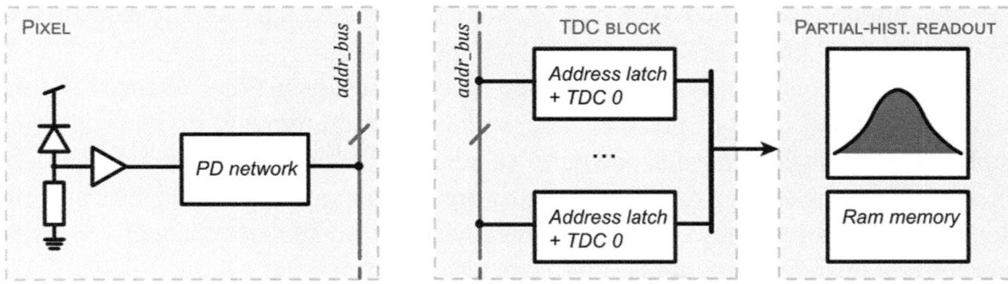

FIGURE 1.32 Simplified block diagram of the concept described in [Zhan19a].

Event-driven TDC triggering

The first SPAD-based implementation that approached an event-driven architecture was reported in [Zhan19a]. This architecture not only marked the first integrated histogramming implementation for a full array, but also generated histogram data in an event-driven manner.

Figure 1.32 illustrates a simplified block diagram of a column in the sensor from [Zhan19a]. The pixel circuitry is reduced to the SPAD front-end and a pull-down network. When a photon is detected, the address of that pixel is written onto a shared bus, which also incorporates a collision detection mechanism. If no collision occurs, the event triggers one of the six available TDCs. These TDCs are dynamically allocated and send the output data to a partial-histogram readout block, which is shared among each of the four half columns. This block can also be configured to integrate photon counts for 2D imaging.

The implementation of the event-driven trigger for the TDC block ensures efficient data throughput from the pixel array to the partial-histogram readout block, as pixels that do not detect photons are not scanned. However, the sensor interface with the final application is still frame-based, potentially limiting the exploitation of the advantages offered by event-based architectures.

Asynchronous readout for Image-Scanning Microscopy

Image-scanning microscopy is a technique used in microscopy to capture high-resolution images of biological samples. Instead of using traditional methods that capture an entire image at once, image-scanning microscopy works by scanning a focused spot of light across the sample in a grid-like pattern. As the spot moves, it captures detailed information at each point, building up the full-resolution image. This allows for high-resolution imaging even in thick samples, as it can selectively focus on different depths. It is particularly useful for studying structures within living cells or tissues.

This specific application justifies the integration of a detector array with a limited number of pixels but with asynchronous readout [Butt20], removing frame rate limitations. In the work presented in [Butt20] a 5 × 5 SPAD array was implemented, with pixels being asynchronously read out. The architecture of the implemented pixel directly transmits off chip the output signal from the SPAD front-end. Each pixel's output is directly connected to an output pad, which is then individually measured to produce the final image.

While this architecture has demonstrated excellent results in Fluorescence Lifetime Imaging [Cast19], it may not meet the scalability requirements of most imaging applications. Using a dedicated pad per pixel restricts the number of pixels that can be integrated into the array. Additionally, under higher photon arrival rates, the transmission of single photons may consume the bandwidth of any readout channel, rendering this concept unsuitable for general-purpose vision sensors.

SPAD vision system with spiking neuromorphic processor

In their work, authors from [Shaw23] introduced a novel SPAD-based vision system integrating a spiking neuromorphic processor. This integration marks a significant advancement in the field of neuromorphic vision. The architecture of the vision sensor is similar to that of [Berk15]. The neuromorphic processor is designed to directly process spiking data, without the need for reconstructing an image. Therefore, the system can exploit the advantages of the spiking nature of SPAD signals.

The key novelty of their approach lies in the integration of the neuromorphic processor and the SPAD-based vision sensor within a single chip. This integration is pivotal as it enhances the overall efficiency of the system. The neuromorphic processor introduces a unique computational paradigm to the vision sensor. It processes address events in a spiking fashion, mimicking the behavior of neurons in biological systems. This innovative integration is capable of processing events at a remarkable rate of up to 1.2 Giga-events per second.

However, it is essential to highlight that while the work in [Shaw23] provides an extensive evaluation of the complete temporal pulses dataset, the experimental results are primarily related to the neuromorphic processor. Specific experimental data regarding the performance and characteristics of the pixel array within the SPAD-based vision sensor are not explicitly detailed in their study. Instead, their emphasis lies predominantly on the capabilities of the neuromorphic processor and its integration with the SPAD-based vision sensor. They use a simulation data set to feed the neuromorphic processor. Nonetheless, it is important to acknowledge this unique aspect of their approach within the broader context of research in the field.

Event-driven coincidence detection for Quantum Microscopy

Quantum Microscopy requires efficient detectors able to identify temporal correlations among photons. Photon coincidences are usually detected by postprocessing their timestamps. This has been traditionally implemented using power-hungry per-pixel TDCs or counting photons in a well-defined temporal window.

The conventional frame-based readout methods come with limitations regarding the duration of observation per frame and tend to produce a substantial amount of redundant data. To address this, the study in [Seve23] proposed a SPAD-based sensor specifically designed for Quantum Microscopy, incorporating an event-driven readout system for photons detected within a defined temporal window.

In Figure 1.33, a simplified diagram of the pixel from [Seve23] is depicted, highlighting two key components in its operation. Firstly, when a photon is detected,

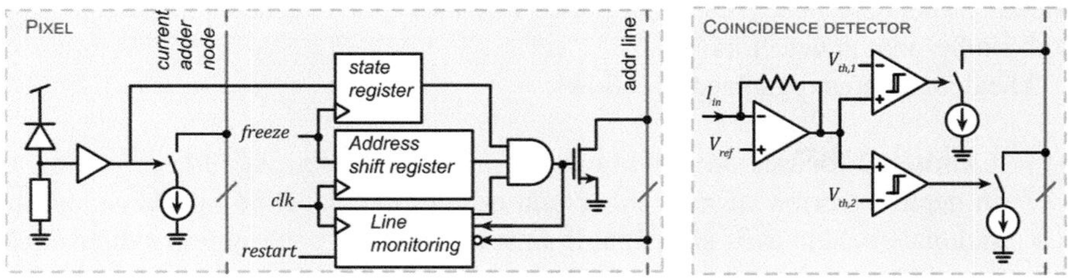

FIGURE 1.33 Simplified block diagram of the sensor reported in [Seve23].

a transistor facilitates the flow of current into a current adder node for a duration equivalent to the dead time of the SPAD. This cumulative current is then directed to a *Transimpedance Amplifier* (TIA), shared among a cluster of pixels. The output of the TIA is subsequently compared to a voltage threshold to identify when two or more photons were detected simultaneously. Note that the output of the coincidence detector is again a current to replicate this operation among clusters. The second crucial element is the readout logic integrated into each pixel. Upon detecting photon coincidences, the *freeze* signal is activated and the state of the pixels (whether a photon was detected or not) is registered. Also, the address shift register is reset with the address of that pixel. Following this, up to three pixels that detected a photon in that period transmit their addresses serially, utilizing a shared transmission line that is normally high state due to a common pull-up device. The pixels monitor the state of the line, and if the transmitted data differs from the pixel address, it identifies a collision and waits for the next cycle to transmit its address.

However, this architecture may not be easily adapted for most applications of image sensors beyond the field of Quantum Microscopy. The serial transmission of data corresponding to individual photons in a high-resolution pixel array could potentially distort the temporal information associated with each event. Also, the definition of the pixel address at the hardware level imposes additional complexities to the design.

1.5 OUTLINE OF THE BOOK

This book is centered on the conception, development, and implementation of an innovative vision sensor, marking the convergence of SPADs and event-driven architecture within a scalable architecture. The book aims to provide alternatives to existing paradigms prevalent in SPAD-based sensors by departing from conventional frame-based architectures. By prioritizing the transmission of meaningful data over extensive data volumes, the proposed sensor system addresses critical challenges in data handling and processing. This transformative approach not only improves storage and processing requirements but also opens new avenues for diverse applications, including but not limited to augmented reality or autonomous driving.

The book also aims to propose architectures that integrate DVS with SPAD detectors, employing event-driven readout schemes. It introduces the concept of discrete DVS, where event detection is initiated by the reception of a specific number of photons. This book describes, models, and validates a novel concept tailored to align with single-photon detectors.

The document is organized as follows:

- **Chapter 2** focuses on describing the proposed camera system, providing an in-depth overview of its architectural design, individual components, and the rationale behind their selection. It aims to underscore the system's distinctiveness in comparison to existing solutions, emphasizing its unique advantages.
- **Chapter 3** delves into the detailed composition and functionalities of the proposed vision sensor. It elaborates on each element constituting the sensor and thoroughly examines their operational behaviors and interactions within the system.
- In **Chapter 4**, the focus shifts towards the exploration of the pixel architecture. This chapter encompasses a study of the different elements comprising the pixel, offering insights into their design, alternative proposals, and potential enhancements for augmenting or extending the pixel's functionality and capabilities.
- **Chapter 5** embarks on a rigorous bottom-up validation process of the entire system. It initiates with an evaluation of SPAD devices and progresses through electrical verifications of the sensor. Notably, this chapter culminates in real-world validation, showcasing the sensor's functionality in both controlled laboratory conditions and practical scenarios. Particularly, it highlights its application in an astronomy-based scenario, accentuating the sensor's event-driven nature, which enables the extraction of information from celestial events with a high temporal resolution, an unprecedented feat in the field.
- Finally, **Chapter 6** encapsulates the findings derived from the study and outlines prospective paths for further enhancements in sensor technology and ongoing research endeavors.

2 A SAER vision system

LiDAR systems have become increasingly prevalent across a wide range of applications, particularly as simultaneous localization and mapping systems have become crucial in areas such as autonomous vehicles [Li20], augmented reality applications [Gene22; Xie23], object tracking [Asva16; Peng23], and autonomous missions and robots [Yang22]. These systems are used to map an unknown environment and to determine the trajectory of the sensor.

With the development of SPAD-based ToF technologies, the development of LiDAR systems has become of increasing interest. However, these systems tend to be costly, power-consuming, and require significant storage capacity and post-processing. Therefore, researchers and companies have been striving to optimize these metrics. Techniques such as on-chip histogram building [Zhan19a] and partial histograms [Tane22], to name a few, have been employed to reduce the energy consumption of the sensor and to reduce the time spent reading out redundant data. However, existing LiDAR sensor architectures are based on sequential readout schemes, whereas the input data of these sensors, i.e., incoming photons, is inherently asynchronous.

This chapter covers the design and development of a *SPAD-based with AER communication* (SAER) Vision system, which represents the first-of-its-kind LiDAR system with asynchronous communication. The core of this system consists of an Event-Driven SPAD-based 3D vision sensor, which significantly enhances the system's performance in terms of power consumption and data throughput optimization.

2.1 OVERALL ARCHITECTURE OF THE SYSTEM

A LiDAR system primarily consists of two essential components: an emitter and a receiver. The former emits laser light, which interacts with objects within the scene. The receiver detects the reflected light and calculates the ToF to generate a 3D image. Depending on the specific implementation of ToF in the LiDAR system, namely direct or indirect, the signal is extracted from a time or phase measurement, respectively. Furthermore, the laser light may be either pulsed or modulated.

As discussed in Section 1.1.3, dToF sensors offer several of advantages, such as long-range operation and high immunity against background illumination. However, dToF sensors require histogram analysis, typically performed outside the pixel,

DOI: 10.1201/9781003427490-2

thereby requiring additional memory and processing resources. On the other hand, iToF sensors require fewer components in the pixels. As a result, large pixel arrays are easier to implement. Therefore, the choice between direct or indirect ToF techniques must be carefully evaluated depending on the application requirements.

2.1.1 System architecture and block diagram

The SAER system is presented as a camera system including all necessary elements to acquire 2D and 3D images. Figure 2.1 depicts the SAER system, providing a visual representation of its key components.

Figure 2.1 shows the block diagram of the proposed LiDAR system. It is composed of the following modules:

- **Optical emitter.** A VCSEL and an EEL are employed to emit light pulses or modulated light.
- **Receiver.** An Event-Driven SPAD-based 3D vision sensor is the core of the system. It senses the reflected light from the emitter and computes the ToF of the received light. It supports multiple operating modes and uses the AER protocol to convey information off-chip.
- ***Power Management Unit* (PMU).** This subsystem generates and regulates the different power supply voltages required by the system, including the high voltage to bias SPADs above breakdown.
- ***Control and Processing Module* (CPM).** This module is implemented in an FPGA, and it is in charge of controlling the overall system and pre-processing the output data of the vision sensor.

The system is intended to be connected to an external host and be controlled using an *ad-hoc* software platform that selects the operation mode and facilitates the representation of output data by reconstructing frames based on the incoming events from the sensor. Nevertheless, the performance of the application can be significantly

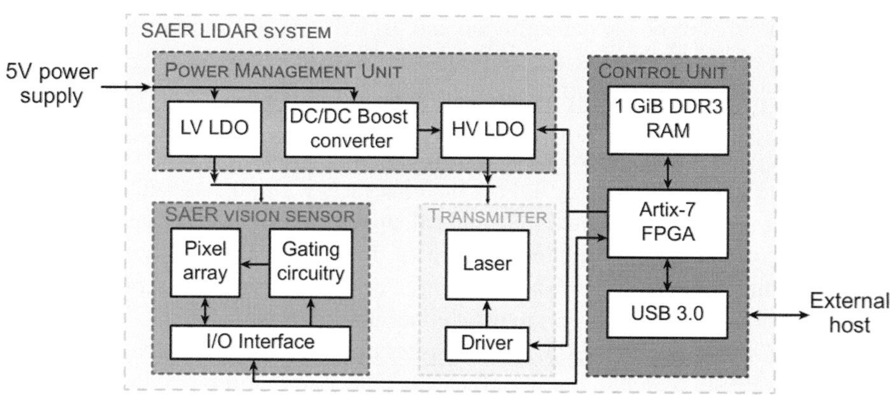

FIGURE 2.1 Block diagram of the SAER LiDAR system.

enhanced through the incorporation of a neuromorphic accelerator that directly leverages AER data for the extraction of scene features and information, since the concept of frame is meaningless in event-driven sensors.

The choice of the optical emitter is often determined by the range requirements and integration feasibility of the application. In addition to factors such as output beam quality, wavelength bandwidth, and cost, VCSELs offer advantages concerning system embedding with minimal area occupation. However, VCSELs tend to have lower power and efficiency compared to alternatives like EELs [Main13]. In this case, a VCSEL was selected as the emitter for ranges of 1–10 meters, which are especially suitable for laboratory testing due to limited space in indoor experimental setups. An EEL was also included in the system to extend the sensing range of the system. Additionally, the SAER system incorporates an SMA output to enable the user to use external laser sources. However, this comes at the expense of additional delays that must be carefully calibrated to ensure optimal system operation.

The SAER system incorporates a 64×64 multi-mode vision sensor [Gome23d], which measures the ToF of the laser pulse obtained following VCSEL emission and reception of the reflected light. This IC, manufactured in a 110-nm process, employs SPADs as single-photon detectors. The main novelty of this device is its event-driven operation, where pixels generate events asynchronously and request access to a shared readout bus. This feature allows for power and processing resource savings, as only meaningful pixels are readout, i.e., pixels that do not detect photons do not convey information. The vision sensor operates as both a 2D sensor for scene illumination detection and a 3D sensor for depth measurement. The intensity information is pre-processed by the CPM and can be used for on-the-fly sensor configuration updates. Further insights into the sensor's operation modes will be presented in Section 2.3.

The vision sensor chip is connected to a PMU to generate, control, and regulate the power of the vision sensor—see Figure 2.2. While most of the circuitry of the system operates at 3.3 V or 1.2 V, SPADs require bias voltages higher than 18 V. From a weight, cost, and complexity point of view, including more than one battery in a system is inefficient. Thus, DC-DC conversion is of special interest. Besides, the efficiency of switched converters can be close to unity [Eric07]. Nonetheless, switched converters are inherently noisy and present an output voltage ripple. Thus, an adjustable *Low-Dropout Regulator* (LDO) regulates the SPAD voltage to reduce power-supply-induced noise in these devices. Further enhancement can be achieved by integrating the PMU directly on chip. This integration offers the advantage of implementing enhanced monitoring circuits that enable continuous adaptation of the SPAD voltage, ensuring insensitivity to PVT variations.

The CPM is implemented in a commercial FPGA integration module for simplicity in the prototyping and debugging stage. Apart from the FPGA, this module includes an external random-access memory and a *Universal Serial Bus* (USB) 3.0 interface that allow for data storage and communication to the host, respectively. While this module is integrated in an independent *Printed Circuit Board* (PCB), the rest of the system is included in the same PCB. Figure 2.2 shows the designed PCB, highlighting the different modules comprising the SAER vision system.

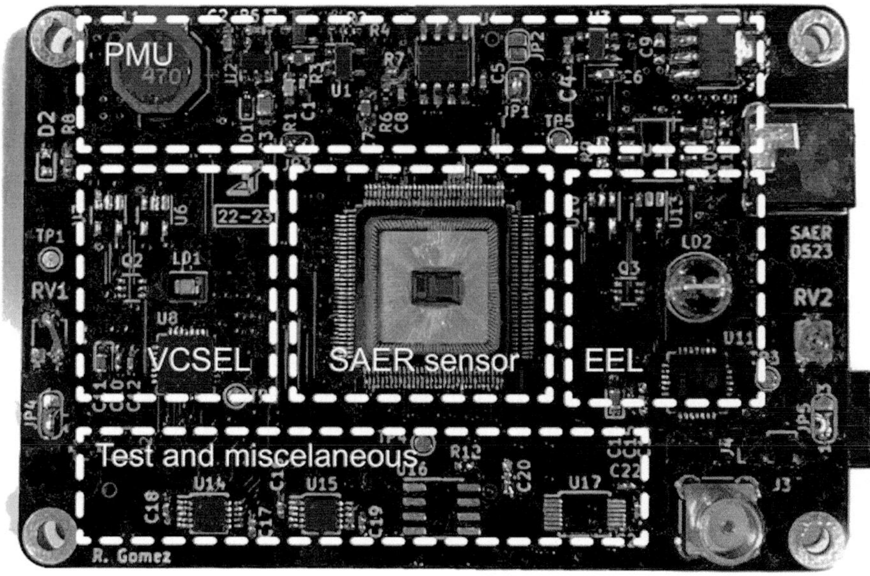

FIGURE 2.2 PCB housing the SAER vision system. CPM is located in an independent PCB connected at the bottom of the main PCB.

2.2 OPTICAL EMITTER

The selection of the optical emitter is one of the most important aspects of a LiDAR system, since its performance considerably limits the performance of the overall system. Therefore, a thorough study of the limitations and characteristics of the emitter, as well as of its advantages and disadvantages, is imperative prior to its integration in the system.

2.2.1 Optical and electrical limitations

The primary limitations of the optical emitter stem from its dynamics and optical power capabilities. Figure 2.3 (a) presents an idealized scenario depicting the emitter's operation. In this scenario, the laser pulse is assumed to be infinitely short, resulting in the emission of all photons simultaneously. Furthermore, the power of the pulse is assumed to be infinite, leading to the emission of all photons simultaneously. Consequently, an infinite number of photons arrive at the receiver with a delay t_{tof}.

On the other hand, Figure 2.3 (b) illustrates a more realistic situation where the laser's energy is spread over time. Consequently, photons are emitted continuously during a duration T_{on}. Although each photon takes the same time to travel from the emitter to the receiver, their reception is temporally shifted, as is their emission. Nonetheless, it remains feasible to compute Δt_{tof} based on the rising or falling edge of the received light profile. Furthermore, the energy of the reflected pulse is lower than that of the original pulse. This energy, i.e., the number of incoming photons,

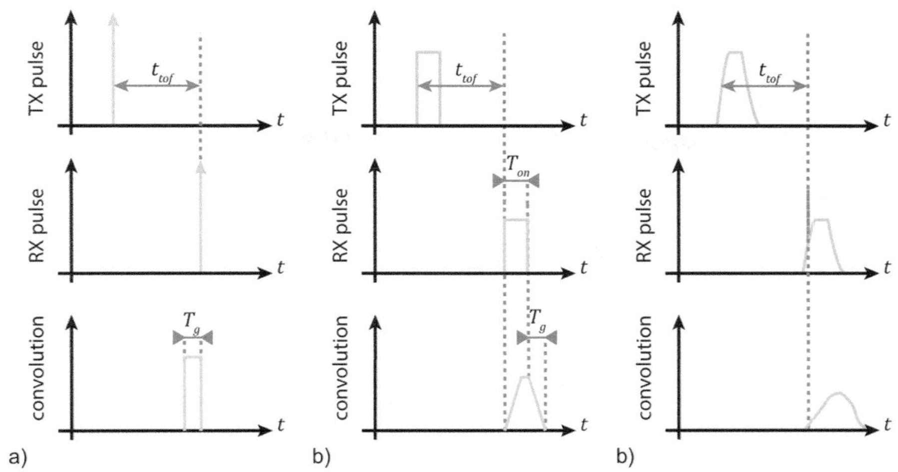

FIGURE 2.3 Transmitted (TX) and received (RX) pulse and convolution with a square signal of width TG of: a) Ideal pulse. b) Non-zero pulse width pulse. c) Real-world pulse.

is limited by the reflectivity of the target and the distance they travel (following the inverse square law). These effects are well-covered by the LiDAR equation [Weit06].

Finally, Figure 2.3 (c) depicts a real-world scenario where the dynamics of the laser precludes making T_{on} short enough, making it challenging to compute t_{tof} accurately. However, it is still possible using the CoM algorithm. This approach increases the computational complexity and restricts the measurement range, as the entire profile, constrained by T_{on}, must fall within the timing range of the measurement, T_{span}, to compute the CoM without error.

The width of the received signal also depends on the method used to compute the ToF. Where time-gating techniques are employed, the output signal of each pixel is the convolution between the time gate and the incoming light profile [Mori20a]. The bottom drawings in Figure 2.3 depict the convolution product between the RX pulse and a square signal with a pulse width equal to T_g. As a direct result of this non-zero pulse width, the duration of the measured pulse increases by T_g. Therefore, the measurable distance range, d_{span}, can be approximated as:

$$d_{span} = \frac{c}{2}\left(T_{span} - T_g - T_{on}\right) \tag{2.1}$$

As previously described, the energy of the received laser pulse is lower than the energy of the original light pulse. This reduction in energy directly affects the number of photons received compared to those initially emitted. The power of a laser pulse diminishes with distance in a manner proportional to the inverse square of the distance traveled. Hence, it is crucial to consider the power and field of view of the laser when determining the operational range of the sensor. These factors must be carefully selected to optimize both power consumption and system performance. It is important to note that since a laser pulse with low energy dissipates before returning to the sensor, it requires a greater number of samples to build histograms with an acceptable SNR. By appropriately addressing these considerations, the efficiency and effectiveness of the system can be enhanced.

At this point, it becomes evident that an ideal laser pulse must:

1) show fast rise and fall times to facilitate ToF computations and the minimization of T_on and
2) enable the emission of a large number of photons within a short duration, i.e., the optical power must be high enough, which is a challenging task.

Optical power is related to electrical power, and hence to current consumption. Therefore, achieving sharp transients requires a rapid increase in current consumption within the laser device, denoted as high *di/dt*. However, this metric is typically constrained by the parasitic inductance in the loop, as illustrated in Figure 2.4. The inductance comprises two primary contributions: the inductance arising from the bonding wires of the component itself, $L_{bonding}$, and the parasitic inductance of the PCB traces, L_{trace}. Commercial products typically report rise and fall times in the range of 1 ns [ams18], which may still be considered high for sub-cm resolution. Various solutions can be employed to mitigate the impact of these parasitics. Firstly, the use of multiple bonding wires in parallel is often implemented during the packaging of these devices to minimize the inductance of the bonding wires, as depicted in Figure 2.4 (a). However, it requires the driver to implement the same technique. Additionally, to further reduce this effect, the driver and the laser device can be assembled within the same package, potentially employing flip-chip techniques.

Note that both the package and the PCB introduce additional parasitic elements such as resistance and capacitance, which impact the bias point and dynamics of the emitter. For more intricate models, references are available in the literature [Hama20].

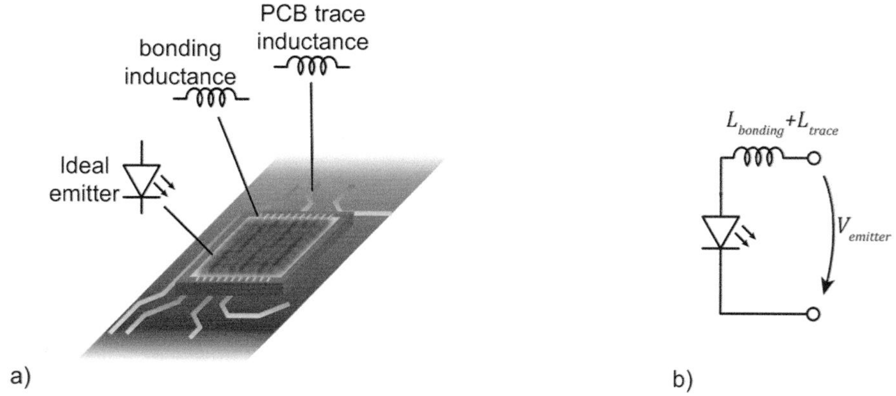

FIGURE 2.4 a) Drawing depicting emitter parasitics within a PCB. b) Equivalent emitter model with included parasitic inductances.

2.2.2 Eye-safety considerations

In laser applications, it is essential to consider eye-safety rules, particularly when humans are involved in the final application. While Section 2.2.1 previously emphasized the importance of high laser power, it is imperative to recognize the significance of complying with established eye safety standards. Industrial applications, where humans are not present, may leverage high-energy light that aligns with the absorption peaks of silicon, typically around 500 nm. However, applications such as autonomous driving or face recognition require a shift towards NIR or even *Shortwave Infrared* (SWIR) wavelengths to ensure sufficient laser power without posing a danger to individuals, even though silicon devices exhibit poor efficiency at these wavelengths.

In order to ensure compliance with eye safety requirements, lasers must adhere to the *IEC 60825-1* international standard [IEC14]. This standard provides guidelines and specifications for the safe use of laser products, considering various parameters including wavelength, optical power, and repetition rate and pulse width in the case of pulsed lasers.

Within the IEC 60825-1 standard, the *Maximum Permissible Exposure* (MPE) limits are defined for different wavelengths. These limits are established based on the potential harm that laser radiation can cause to the human eye and skin. By following the standard, laser systems can be designed and operated in a manner that ensures the safety of individuals. The MPE limits take into consideration factors such as the sensitivity of the human eye at different wavelengths, the potential for thermal damage to ocular tissues, and the body's natural aversion response to bright light. By adhering to these limits, laser systems can operate within safe parameters and minimize the risk of eye damage or injury to individuals who may come into contact with the laser radiation. The standard establishes different constraints for the MPE depending on the nature of the laser light, which can be classified into continuous-wave light or pulsed light.

The selection of maximum power, repetition rate, and pulse width of the laser depends on the specific wavelength being used. The standard provides tables and calculations to determine the appropriate limits for each parameter based on the wavelength range. For instance, shorter wavelengths in the ultraviolet and visible spectrum show lower MPE limits, while longer wavelengths in the NIR and SWIR range have higher limits. To provide a brief explanation, the maximum MPE of a repetitively pulsed laser needs to be restricted based on the most restrictive of the following requirements:

1) The exposure of a single pulse must not exceed the MPE indicated in the tabulated data.
2) The average exposure of a pulse train with a duration of T must not exceed the MPE of a single pulse with the same duration, T.
3) The average exposure of pulses must not exceed the MPE multiplied by a correction factor dependent on the number of pulses within an exposure.

This limitation is particularly relevant for LiDAR applications. On one hand, it is desirable to distribute the laser energy as effectively as possible, especially for long distances (hundreds of meters). On the other hand, increasing the repetition frequency as much as possible is also desirable to achieve lower latency in constructing histograms.

Furthermore, IEC 60825-1 establishes limits to classify lasers into different classes. For a laser to be categorized as Class 1, signifying safety under all circumstances, the energy of the pulses must not exceed the Accessible Emission Limit. The criteria mentioned earlier for repetitively pulsed lasers also apply for this classification.

In conclusion, the IEC 60825-1 standard provides clear guidelines for selecting maximum power, repetition rate, and pulse width of lasers based on the wavelength being used. Adhering to this standard ensures that laser applications, especially those involving human interaction, are conducted in a safe and responsible manner, prioritizing the well-being of individuals.

2.2.3 Optical emitter selection

As discussed in Section 1.3.2, the choice of emitter technology plays a crucial role in the performance of a LiDAR system. There are several factors impacting on the selection process, such as power efficiency, beam quality, and integration suitability.

VCSELs offer a reasonable peak power with excellent beam quality, making them an ideal choice for short-range applications such as face recognition or AR. Their compact size aligns well with the goal of creating a highly integrated LiDAR system. However, it is important to note that VCSELs may have limitations in terms of power efficiency, which may make them less suitable for long-range applications.

EELs are preferred for medium and long-range applications due to their typically higher power efficiency. Most commercial EEL solutions offer rise times in the order of 1 nanosecond, making them well-suited for a variety of scenarios. However, they may require additional optical components to enhance beam quality, and their duty cycle is limited (usually to 0.1%), primarily due to thermal considerations.

To provide a versatile system for short-range and mid-range applications, the SAER vision system incorporates two laser modules. The first is a VCSEL (V205 from ams OSRAM) with a peak power of 2 W at 940 nm, which is particularly well-suited for short-range distances, making it an excellent fit for laboratory settings. The second module is an EEL (SPL PL90 from ams OSRAM) with a peak power of 25 W at 905 nm, making it the preferred choice for mid-range applications. These decisions were driven by several factors. Firstly, the limited operating range of VCSELS is not a significant drawback for a proof-of-concept prototype. Furthermore, the integration potential of VCSELs may offer opportunities for the future, for instance the camera system may incorporate additional functionalities, leveraging the addressability of VCSEL arrays [Chen23] or their beam steering capabilities [Pan19]. These advantages contribute to the overall size reduction and improved performance of our LiDAR technology. On the other hand, the inclusion of an EEL allows increasing the depth range of the camera system, to offer a more versatile solution for ranges where the reflected light pulse of the VCSEL is attenuated.

2.2.4 Driver Design

In order to achieve a proper light profile and ensure optimal performance of the optical emitter, it is essential not only to select an appropriate emitter but also to design a suitable driver. The driver must deliver the current required to support operation at the targeted frequency, and to achieve fast rise and fall times for precise control of the emitted light.

Figure 2.5 (a) illustrates the simplest driver configuration, employing a MOSFET as an analog switch [Vixa19]. However, this basic scheme has certain limitations and disadvantages. One significant drawback arises from the presence of parasitic inductances in both the MOSFET and the laser itself. These parasitic inductances introduce undesired effects that can degrade the dynamic performance of the driver, as introduced in Section 2.2.1. Additionally, when the laser is switched off, the stored energy in the parasitic inductance can lead to ringing, as depicted in Figure 2.5 (b), causing potential damage to the connected devices. These issues restrict both the rising and falling edges of the laser pulse, as well as the minimum pulse width that can be achieved.

To overcome these challenges, more sophisticated driver circuits can be employed. Figure 2.6 illustrates an enhanced driver scheme that addresses these limitations

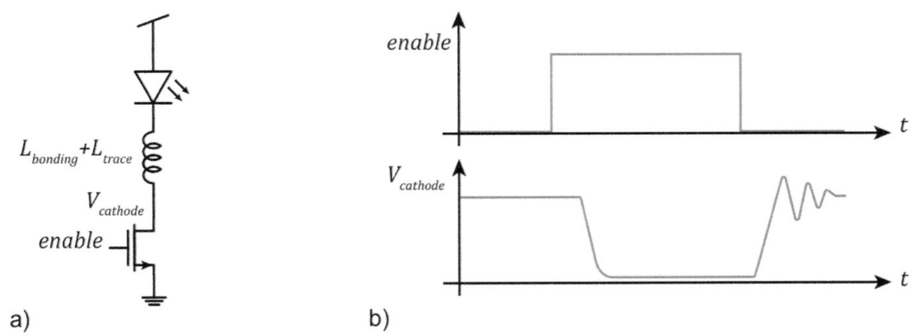

FIGURE 2.5 a) Emitter driver composed of a single NMOS. b) Waveform diagram.

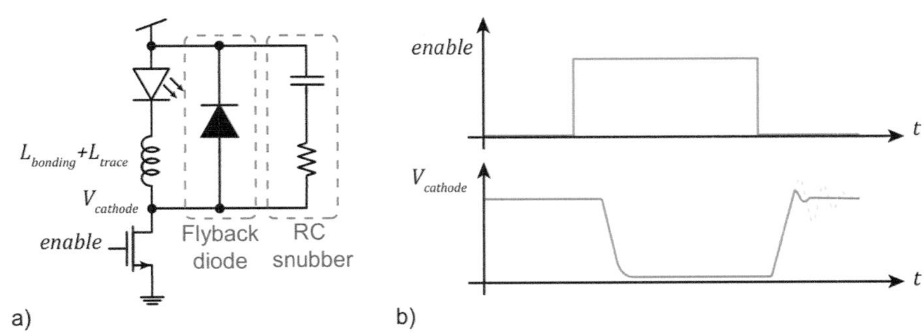

FIGURE 2.6 a) Advanced emitted driver, including flyback diode and RC snubber. b) Waveform diagram.

[icH19]. In this design, a flyback diode is included in parallel with the laser, providing a safe current path when switching off the laser. This helps mitigate the effects of parasitic inductances and prevents excessive voltage transients. Furthermore, an RC snubber circuit is introduced to account for any residual oscillations and dampen them effectively. By incorporating these additional components, the driver can deliver improved performance, reducing ringing and enabling better control over the emitted light.

However, driver dynamics may also be limited by the input impedance of the power MOSFET. If a high-power laser is required, the ON resistance of the device must be low enough. However, this usually comes at the cost of a larger gate capacitance, which complicates the design of the driver stage, even constraining the rise and fall time of the input signal. Thus, a power-speed trade-off may appear in the design of the driver.

Furthermore, the power of the laser can be modulated by adjusting the strength of the transistor. This can be done by simply adjusting the amplitude of *enable.* In fact, a combination of two transistors is usually employed, the first acting as a current source that limits the output current and the latter operating as a switch [icH19].

2.3 THE SAER VISION SENSOR

The SAER vision [Gome23d] sensor serves as a fundamental component within the SAER vision system and differs from conventional LiDAR systems in the usage of an event-based asynchronous readout mechanism. This unique feature empowers the sensor to govern data transmission without the continuous readout of irrelevant pixels being needed. Additionally, the pixels of the SAER vision sensor offer the capability of online sensitivity adjustment, making it well-suited for the implementation of advanced algorithms, HDR imaging, and adaptability to varying ambient conditions. The design of the sensor and its functionalities play a critical role in enhancing the overall performance and flexibility of the SAER vision system.

2.3.1 DESCRIPTION OF THE SENSOR

Figure 2.7 presents the simplified block diagram of the SAER sensor. The sensor incorporates a 64×64 array of SPAD-based intelligent pixels. These pixels are designed to function similarly to cells (metaphorically called neurons) found in octopus retinas [Culu03]. However, the core component of these SAER cells is a SPAD, capable of detecting individual photons. Furthermore, these cells can be individually programmed to generate events with varying sensitivity to incoming photons. The in-pixel circuitry used for controlling this sensitivity can also serve to implement in-pixel pre-integration, reducing data generation in different operation modes. Additionally, SAER pixels can be independently disabled, enabling adaptive resolution reduction or the implementation of foveated vision algorithms [Wang05] based on the spiking data of an active neuron.

The operation of the neuron array is governed by the *enable* signal, enabling precise modulation of their activity. To shape this signal effectively, the SAER

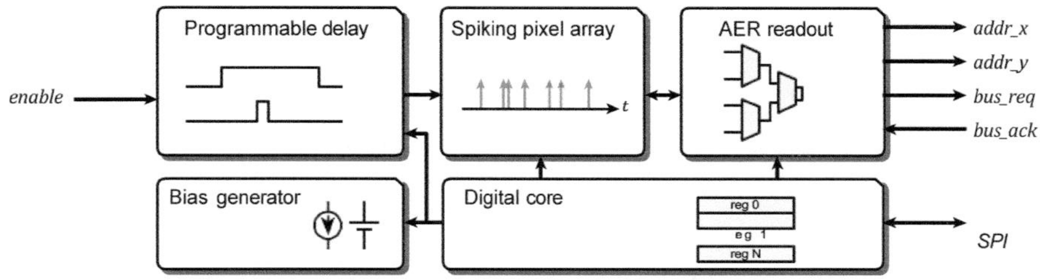

FIGURE 2.7 Simplified block diagram of the SAER vision sensor.

sensor incorporates a Programmable Delay block primarily utilized for ToF detection. This block introduces a programmable delay with a resolution lower than 1 ns. Additionally, specialized circuitry facilitates pulse width programming. This delay and shaping operation enables the measurement of the timing profile of a correlated light pulse. Consequently, ToF can be computed using a CoM algorithm.

The AER readout block serves as the intermediary between the neuron array and the external receiver. Its role encompasses reading the events generated by the neurons and arbitrating access to the readout bus for the pixels. This block implements the AER protocol through a word-serial link, where the output data correspond to the x- and y- coordinates of the pixel responsible for generating the event, referred to as *addr_x* and *addr_y*, respectively. The signal *bus_req* indicates the presence of valid data at the bus address, while the signal *bus_ack* informs the sensor that the receiver has successfully stored the address. This, in turn, allows the AER readout block to initiate the resetting of the pixel that initiated the event.

To facilitate integration and precise control, both the programmable delay block and sensor configurations can be controlled using a *Serial Peripheral Interface* (SPI) that communicates with the digital core of the SAER sensor. This interface enables efficient communication and management of the operation of the sensor, providing programmability and optimization options to meet specific application requirements.

Although primarily designed to work as an octopus retina, where pixels continuously transmit spike-encoded information in a free-running operation, the SAER sensor offers versatility by supporting various operation modes based on the sequence of operation signals and event processing. These operation modes include:

- ***Free Running*** **(FR)**: In this mode, pixels encode the intensity level into frequency, continuously transmitting pulses to CPM. This mode allows for real-time, continuous data transmission and processing.
- ***Time-to-N-Photons*** **(TNP)**. Pixels generate a single event when they detect a specified number of photons, N, after the *enable* signal is asserted. This mode is similar to *time-to-first-spike* operation commonly found in conventional asynchronous imagers [Culu03; Gome23a]. It enables precise timing measurements and control on the shot noise impact on pixel's data.
- ***Quanta Imaging*** **(QI)**. In this mode, the illumination level is encoded into the probability of detecting at least k photons within a given time frame. By setting

k to 1, a naive integration can be performed at the pixel level, reducing data throughput, and increasing the equivalent frame rate. This mode is particularly useful for applications where data reduction and high-speed imaging are desired.

- ***Direct Time of Flight* (dToF)**. The sensor computes the ToF of a laser pulse by shifting the enable signal with respect to the laser pulse. This mode is well-suited for long-range applications. The effective range is limited by the power of the optical emitter employed.

By offering a range of operating modes, the SAER sensor exhibits remarkable versatility, enabling adaptation to diverse imaging scenarios and facilitating the extraction of valuable visual information based on specific application requirements.

2.3.2 Events in the SAER vision sensor

The meaning of an *event* may vary depending on the specific sensor and its intended application. For instance, in the case of a DVS, an event typically represents temporal contrast, indicating a variation in light intensity at a particular pixel over time. However, in other sensor types, an event might signify spatial contrast, such as edge detection. It is worth noting that the interpretation of events can differ at the algorithmic level, where a sequence of events can be analyzed to detect patterns or extract valuable insights about the observed scene.

In the SAER vision sensor, the spiking neurons generate spikes when a predefined number of photons are detected. The programmable nature of the sensor allows individual adjustment of this threshold in each pixel, facilitating the extraction of relevant information. Additionally, the event generation can be further adjusted based on the timing of control signals, offering adaptability to various scenarios. From a statistical perspective, an event can be understood as the moment when the signal of a pixel reaches a defined SNR from a photon shot noise point of view [Jane07]. This aspect can be particularly interesting. Assuming that the number of photons is set sufficiently high to make shot noise the primary noise source, we can estimate the ideal frequency deviation of the spiking signals, which is influenced by the inherent quantized nature of light.

This event-driven approach provides a unique framework for smart visual processing in the field of SPAD-based sensors, enabling efficient scene analysis and supporting advanced algorithms for a wide range of applications. Figure 2.8 (a) depicts the essential operations on which the SAER vision system relies: photon integration and event classification. The former is implemented in the vision sensor, while the latter is performed by the CPM in the current implementation. Nevertheless, event classification can be performed at the pixel level, as will be discussed in Chapter 4.

Figure 2.8 (b) provides a visual representation of event generation in the SAER vision sensor. Photons arrive at the detector in a stochastic and asynchronous manner, following Poisson statistics. The inter-arrival time between N_{ph} Poisson events, given an average photon arrival rate, ϕ, follows a Gamma distribution characterized by shape factor N_{ph} and scale factor $1/\phi$ [Blit14]. This means that, if events correspond

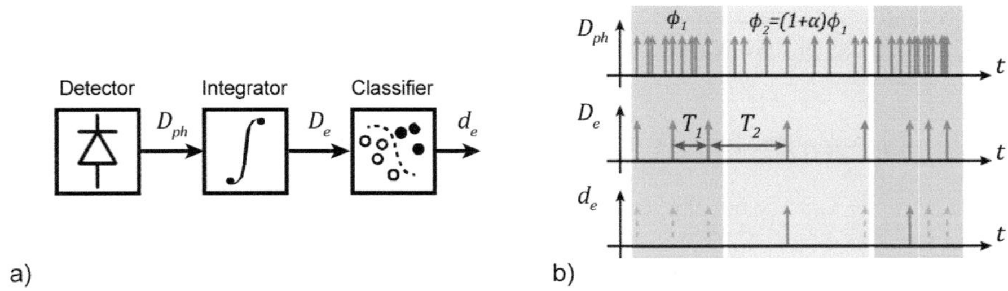

FIGURE 2.8 a) Conceptual block diagram of the basic operations of the SAER vision system. b) Waveforms corresponding to photon arrivals, photon integration, and classification.

to the integration of N_{ph} photons, the parameter ϕ is encoded in the inter-arrival time between two events, denoted as T. It is worth noting that T is a statistical variable with mean and standard deviation values defined as:

$$\mu_T = \frac{N_{ph}}{\phi} \tag{2.2}$$

$$\sigma_T = \frac{\sqrt{N_{ph}}}{\phi} \tag{2.3}$$

Similar to the operation of an octopus retina, these events encode not only information about intensity, but also carry valuable temporal information. This temporal data can be leveraged for tasks like classifying events into intensity events and motion events. For instance, if the photon arrival rate changes from ϕ_1 to ϕ_2, variations in T can be detected, enabling the transmission of only those events related to motion variations. This concept will be explored further in the following section. It is important to highlight that these processing stages lead to a reduction in data generation, as the focus shifts towards relevant information rather than raw data. For example, the data flow pertaining to photons, denoted as D_{ph}, is much larger than that of events, denoted as D_e. Additionally, motion-filtered events, represented as d_e further reduce the data sent to the final application. i.e., $d_e \ll D_e$.

In conventional SPAD-based sensors, dynamic range is constrained by the bit depth of the counters used for accumulating photon counts. The SAER vision sensor, however, overcomes this limitation by employing event-based asynchronous readout.

By encoding information in the frequency of events rather than the absolute photon count, the SAER sensor achieves a substantially wider dynamic range and faster data rates. This is because dynamic range is determined by the resolution of the circuit that records the timestamp of each event, while noise is dependent on the number of photons used for estimating the intensity value. Consequently, employing multiple events for T estimation enables a virtual increase in N_{ph} value and a reduction in noise, at the expense of a reduction in speed. This capability positions it favorably for applications demanding both HDR and real-time processing, also reducing circuit requirements at the pixel level.

2.3.3 Discrete-Dynamic Vision Sensor Implementation

As previously discussed, the inherent digital nature of SPADs and the event-based characteristics of the SAER vision sensor, which involve the collection of a specific number of photons, enable the estimation of the variation in inter-spike time. This capability offers the opportunity to develop a discrete DVS, which considers the inter-arrival time between successive spikes, ΔT, to detect variations in the visual scene. This operation can be executed within the CPM with a marginal increase in latency. However, it is important to note that the bandwidth consumption of the AER readout channel is greater compared to true DVS sensors. This is because the CPM receives a larger number of events than those associated with intensity variations. Nevertheless, the concepts elucidated in this section can be implemented at the pixel level or even in a processing layer if 3D stacking processes are used, to reduce any possible bottleneck associated with the AER readout channel.

Conventional DVS sensors are designed to detect variations of a voltage that logarithmically depends on the photocurrent, I_{ph}, of a linear-mode photodiode [Lich08]. Therefore, temporal variations of this logarithmic dependent variable correspond to relative variations of I_{ph}, as demonstrated by the following expression:

$$\frac{d}{dt}\left(\text{In}\left(I_{ph}\right)\right)=\frac{1\cdot dI_{ph}}{I_{ph}\cdot dt} \tag{2.4}$$

Thus, variations on this voltage can be compared to a voltage threshold that does not require adjustment depending on the absolute value of I_{ph}.

However, single-photon detectors are designed to yield a measurable signal upon the arrival of a single photon. In the case of SPADs, this signal manifests as a digital pulse related to photon detection. While this digital approach aligns well with the quantized nature of photons, the random arrival of these light particles introduces challenges for implementing motion detection at the single-photon level, primarily due to the low SNR. Therefore, more than one photon must be used in the detection.

Since SPAD devices encode the average photon arrival rate, ϕ, in the frequency domain, two main methods are available. The first involves counting photon detections within a specific time interval. However, this synchronous operation introduces limitations in latency, constrained by the exposure time. Additionally, it yields an unknown, photon-flux-dependent, and non-uniform SNR across pixels due to photon shot noise, leading to more false detections in regions with lower illumination. The second method entails measuring the time between the inter-arrival of N_{ph} photons, as in the case of the operation of the SAER vision sensor. Consequently, the variability of this inter-arrival time adheres to Poisson statistics, yielding a SNR of $\sqrt{N_{ph}}$, regardless of the intensity level. Although this can result in extended inter-measurement times and low temporal response in darker regions, it enables continuous operation, well-defined statistics, and independent pixel operation.

Another limitation that appears when implementing a DVS using SPADs is related to the limitations of implementing a logarithmic dependence in the digital domain. Alternatively, a dynamic threshold dependent on the photon counts can be implemented to measure relative variations [Rocc20]. The relative variation or temporal contrast, α, can be defined as:

$$\alpha = \frac{\Delta\phi}{\phi} \tag{2.5}$$

where $\Delta\phi$ represents the variation in the absolute photon flux. To meet the condition for temporal variations to occur, the absolute value of α must exceed α_{th}, being α_{th} the desired contrast threshold. Accordingly, the operation can be redefined as:

$$|\Delta\phi| > \alpha_{th}\phi \tag{2.6}$$

This operation can be implemented whenever an event is generated in the SAER vision sensor, using the inter-arrival time of events to estimate ϕ. Defining T_2 and T_1 as the former and the latter inter-arrival times, α can be estimated as:

$$\alpha = \frac{1/T_2 - 1/T_1}{1/T_1} = \frac{T_1}{T_2} - 1 \tag{2.7}$$

Thus, the condition determining contrast detection can be expressed as:

$$|T_1 - T_2| > \alpha_{th} T_2 \tag{2.8}$$

Note that this operation can be implemented easily in digital hardware when α_{th} is a power of two, since it only implies shifting bits connections, especially if α_{th} is not to be modified.

Although this operation can be implemented at the pixel level, the CPM is the module in charge of classifying events in the current implementation of the SAER vision system. Nevertheless, future implementations may leverage the advantages of advanced technology nodes to increase the functionalities implemented in the pixel level and reduce even more the data throughput at the sensor level.

An important consideration is that T_1 and T_2 are random variables, making this detection mechanism susceptible to false detections, also with the possibility that not all variations will be detected. Therefore, statistical study and modeling are essential to extract all the specifications of a system employing this dynamic vision approach. Given the known event statistics, which are solely dependent on the number of photons defining an event, it is possible to determine the probabilities of correct and incorrect detections, as well as to estimate the average time required for detection or the frequency of false detections.

Probability of detection

The initial step in examining the statistical parameters of this dynamic vision implementation involves recognizing that a Gamma distribution can be approximated as a Gaussian distribution when the shape factor, i.e., N_{ph}, is sufficiently high [Blit14].

Figure 2.9 illustrates the probability density function of Gaussian and Gamma distributions with the same mean and standard deviation for various N_{ph} values. As shown, the error for the case of N_{ph} = 128 (the maximum available in pixels of the SAER vision sensor) is nearly negligible. Hence, for the sake of simplicity, we can consider that inter-arrival times of events follow a Gaussian distribution.

The first parameter of interest is the probability of detection. Figure 2.10 represents the probability density function for two values of ϕ, such as $\phi_2 = (1 + \alpha)\phi_1$, corresponding to an increment in the intensity value. When $\alpha \gg \alpha_{th}$, equation (2.8) will consistently result in a true condition, regardless of the sampled values of

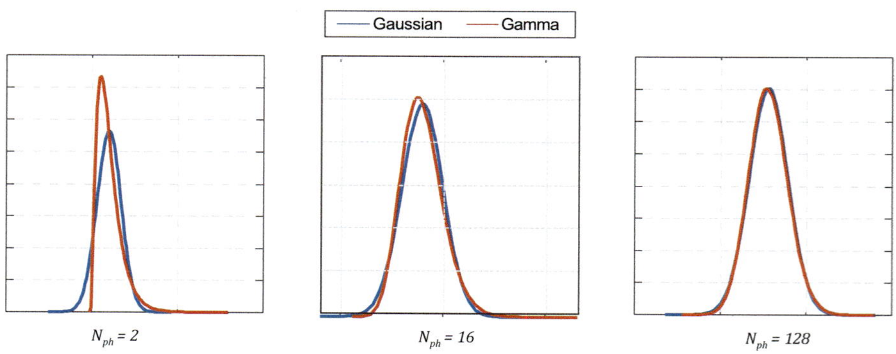

FIGURE 2.9 Probability density function of gamma and Gaussian distributions, for different shape factors.

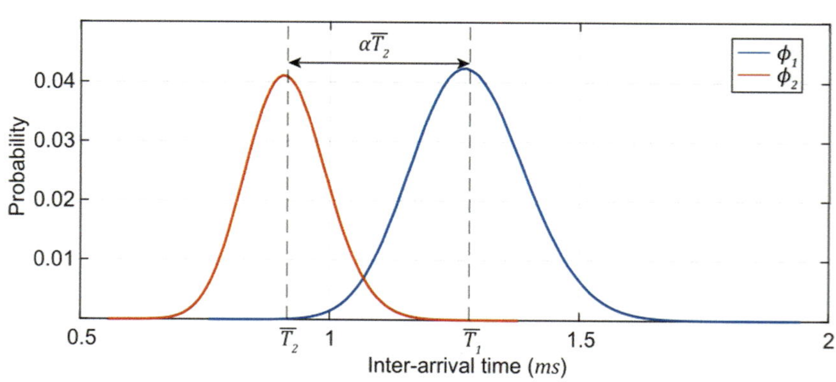

FIGURE 2.10 Probability density function of inter-arrival time for two different intensity conditions.

T_1 and T_2. However, when $\alpha \approx \alpha_{th}$, the scenario is quite different. Three situations may occur, from best to worst scenario:

1. The condition is correctly detected, generating a motion event.
2. The estimated value of α yields a lower value than the actual one, resulting in a failure to detect the motion.
3. The sampled value of T_1 is greater than that of T_2, leading to a detection in the wrong direction.

Therefore, the probability of detecting a positive intensity variation, $P_{det,}$, is a function of α and α_{th}, and can be expressed as:

$$P_{det,p}\left(\alpha,\alpha_{th}\right) = P\left[\hat{\alpha} > \alpha_{th}\right] = P\left[T_1 - T_2\left(1+\alpha_{th}\right) > 0\right] = P\left[X > 0\right] \tag{2.9}$$

Considering the relation between T_1 and T_2, the mean. the variance and SNR of X can be expressed as:

$$\mu_X - \frac{N_{ph}\alpha - \alpha_{th}}{\phi_1 A\, 1+\alpha} \tag{2.10}$$

$$\sigma_X^2 = \frac{N_{ph}}{\left(\phi_1 A\right)^2}\left|1+\left(\frac{1+\alpha_{th}}{1+\alpha}\right)^2\right| \tag{2.11}$$

$$SNR_X = \frac{\sqrt{N_{ph}}\left(\alpha-\alpha_{th}\right)}{\sqrt{\left(1+\alpha\right)^2+\left(1+\alpha_{th}\right)^2}} \tag{2.12}$$

where A is the area of the photodetector.

The same applies for the probability of detecting a negative variation of intensity, $P_{det,n}$. However, in this case, α and α_{th} take negative values, expressing $P_{det,}$ as:

$$P_{det,p}\left(\alpha,\alpha_{th}\right) = P\left[\alpha > \alpha_{th}\right] = P\left[T_1 - T_2\left(1+\alpha_{th}\right) < 0\right] = P\left[X < 0\right] \tag{2.13}$$

Note that the SNR is higher for negative variations. This is because, assuming the absolute value of α_{th} is the same for both cases, the denominator in equation (2.12) decreases for negative values of α and increases for positive ones, while the numerator shows the same tendency for both positive and negative values. Hence, the response is asymmetrical, with a higher likelihood of detecting negative events.

Once the statistical parameters contributing to the operation are known, the probability of detecting variation can be expressed as:

$$P_{det} = \int \frac{1 \quad e^{(1\cdot x-\mu)}dx}{\sigma\sqrt{2\pi}} \tag{2.14}$$

whose integration limits depends on whether the event is positive (from zero to infinity) or negative (from minus infinity to zero).

Figure 2.11 represents the probability of detecting an intensity variation for different values of α, for $|\alpha_{th}| = 0.25$ and $N_{ph} = 128$. The steeper response for negative α values is evident. Additionally, P_{det} crosses 50% at α_{th}, the commonly used value for defining the contrast threshold [Niwa23]. This behavior can be extracted from equation (2.10), as the value of α where $[X > 0] = 0.5$ is the one that makes μ_X equal to 0.

Figure 2.12 (a) and (b) illustrate the variations in $P_{det,p}$ with respect to the number of triggering photons and α_{th}. As previously mentioned, α_{th} sets the contrast threshold, while N_{ph} influences the steepness of the curve. Ideally, a square response is desired. However, this poses a clear drawback for single-photon detectors, as linear-mode

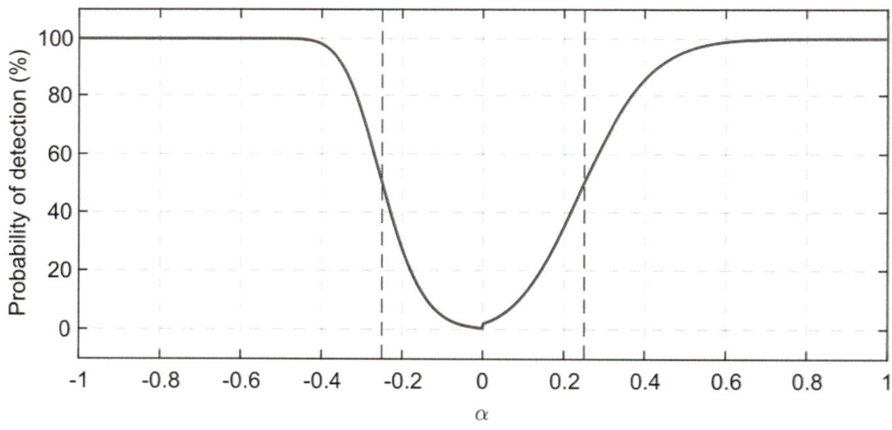

FIGURE 2.11 Probability of detecting an intensity variation with a relative variation of α_{TH} for $\alpha_{TH} = 0.25$ and $N_{PH} = 128$.

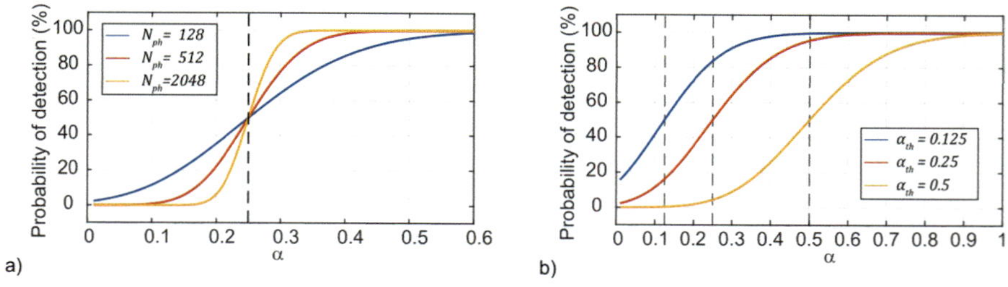

FIGURE 2.12 Probability of positive detection as a function of: a) the number of collected photons ($\alpha_{TH} = 0.25$), b) the contrast threshold α_{TH} ($N_{PH} = 128$).

photodiodes can gather more photons within the same time, providing inherently better performance.

Latency

Given that the estimation of α involves a stochastic process, detecting variations in photon flux may require more than one sample of T, contingent on the actual value of α. Consequently, determining the number of samples, n, required for detection follows a geometric distribution, where the average value of n equals $1/P_{det}$.

Additionally, it is crucial to account for the fact that changes in photon flux may not occur immediately after the last comparison. In fact, the worst-case scenario arises when such variation occurs after detecting $N_{ph}/2$ photons, leading to the averaging of the previous and current inter-arrival times. Thus, expressing the average time needed to detect an event, tdet, takes the form:

$$\overline{t}_{det} = \left(n + \frac{1}{2}\right)\frac{N_{ph}}{\phi} = \left(\frac{1}{P_{det}} + \frac{1}{2}\right)\frac{N_{ph}}{\phi} \tag{2.15}$$

A clear trade-off between contrast sensitivity and latency is evident and can be managed by appropriately adjusting N_{ph} and α_{th}. However, as will be explained in the following section, this also leads to an increase in background noise.

Background activity noise

Background activity noise is usually defined as the frequency of spurious events generated by a DVS sensor [Guo23b]. However, when it comes to single-photon detectors, it may be more appropriate to talk about the probability of false detections. Indeed, we can distinguish between false detections (those not related to intensity variations) and wrong detections (positive events classified as negative events, or vice versa).

The probability of false detections, P_{false}, is a particular case of equation (2.14) when $\alpha = 0$. Therefore, P_{false} can be expressed as:

$$P_{false} = P_{det,n}\left(0, \alpha_{th}\right) + P_{det,p}\left(0, \alpha_{th}\right) \tag{2.16}$$

which clearly depends on N_{ph} and α_{th}, as depicted in Figure 2.12. Thus, to achieve a low background noise level, it is necessary to increase any of these parameters, either penalizing speed or contrast sensitivity.

Figure 2.13 shows P_{false} as a function of α_{th}. From this set of curves, it can be demonstrated that to achieve low contrast sensitivity, N_{ph} must be sufficiently large to minimize the probability of false detections, even reducing it to negligible levels. It is important to note that spatial filtering can also be applied in a subsequent processing stage.

On the other hand, the probability of wrong detections, P_{false} is a particular case of $P_{det,}$ ($P_{det,}$) when α is negative (positive). Figure 2.14 illustrates P_{false} as a function of

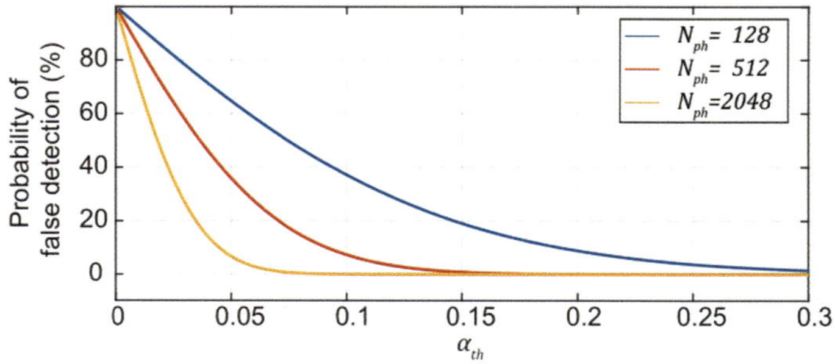

FIGURE 2.13 Probability of false detection as a function of α_{TH}.

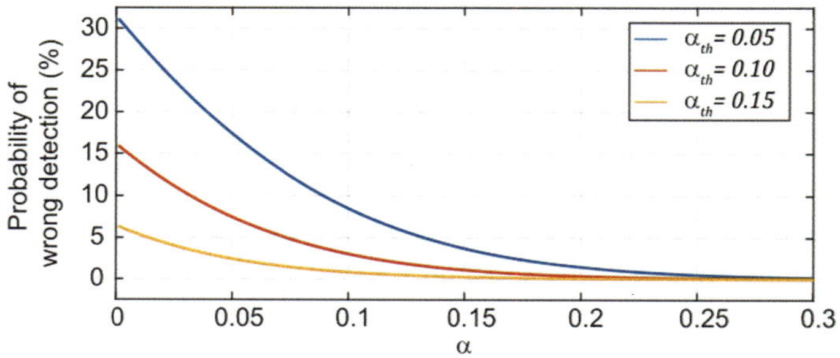

FIGURE 2.14 Probability of wrong detection as a function of $\boldsymbol{\alpha}$, for $N_{PH} = 128$.

α for different values of α_{th}. In this example, the probability of wrong detection above $\alpha = \alpha_{th}$ is lower than 5%, while the probability of right detection is 50%.

2.3.4 Sensor adaptability to the scene

Sensor adaptability to the scene is a key aspect of the SAER vision sensor, enabling it to effectively respond to the characteristics and dynamics of the observed environment. The inherent flexibility of the sensor allows for adaptability in multiple aspects, providing a versatile platform for capturing and processing visual information.

One significant aspect of the SAER vision sensor is the ability to individually adjust the in-pixel photon threshold. This feature allows the sensor to dynamically control the generation of events based on the detected photon count in each pixel. Therefore, the spiking frequency of each pixel can be adaptively updated, optimizing the usage of the shared readout channel. This adaptability ensures efficient utilization of resources, preventing issues such as low event rates that may lead to prolonged periods without information or high event rates that could result in channel saturation.

The adaptability of the SAER vision sensor extends beyond threshold adjustment. The ability of the sensor to capture events and transmit them asynchronously provides

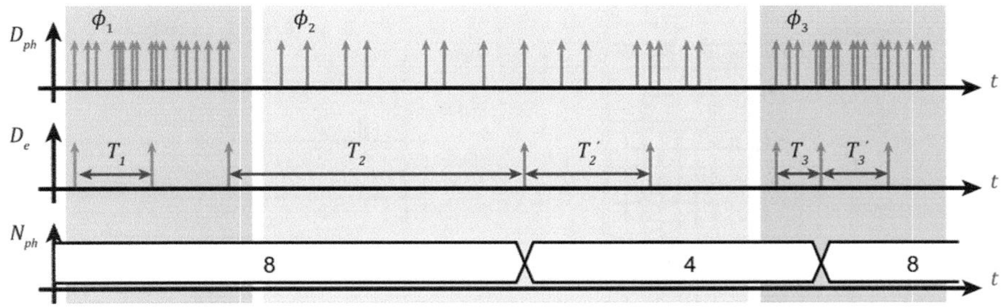

FIGURE 2.15 Concept of vent rate adaptability by adjusting pixel sensitivity.

inherent adaptability to changing scene conditions. The event-driven nature of the sensor allows it to respond in real-time to relevant visual stimuli. Since the external receiver can adjust the pixel sensitivity on the fly during event readout, even access of the readout channel can be guaranteed. To achieve this even access, all pixels must generate events at a similar rate. Hence, to optimize the usage of the readout channel, the goal is that the event rate of each pixel is adapted to be approximately the bandwidth of the readout channel divided by the number of pixels.

Let us assume that a neuromorphic processor reads out the information from the sensor. This neuromorphic processor can process spiking data and benefits from this asynchronous operation. Since information is encoded in the time domain, and thus, the processor monitors the event rate, it can adapt the sensitivity of the pixels once it detects that the event rate varies from the target rate. Figure 2.15 illustrates this concept. As the photon rate varies from ϕ_1 to ϕ_2, so does the inter-arrival time of events, denoted as T_1 and T_2. During readout, the receiver detects this condition and varies the number of photons required to trigger an event, N_{ph}. At this moment, the inter-arrival time varies from T_2 and T' to increase the event rate and compensate for the drop in the photon rate. The same applies in the opposite direction when the photon rate increases to T_3. Note that the adaptation process may take more than one event and depends on the algorithm implemented to adjust the value of N_{ph}. Also, N_{ph} can be programmed as a power of two, ranging from 1 to 128, limiting the accuracy of the algorithm when trying to fix a determined event rate.

Although the current implementation emulates the neuromorphic processor using the CPM, future work may explore the integration of dedicated neuromorphic processors. This would enable an efficient processing of spiking data to properly distribute the bandwidth of the readout channel. By analyzing the significance and relevance of captured events, the sensor can make decisions to optimize data transfer and processing, focusing resources on areas with higher activity or scene importance.

In summary, the adaptability of the SAER vision sensor to the scene is a fundamental characteristic that enables efficient capture and processing of visual information. The ability to individually adjust in-pixel thresholds and the potential for online adaptation offer flexibility in resource allocation and data transmission. By leveraging these adaptive features, the sensor can effectively respond to the dynamics of the observed environment, making it a valuable tool for various applications in the field of SPAD-based imaging.

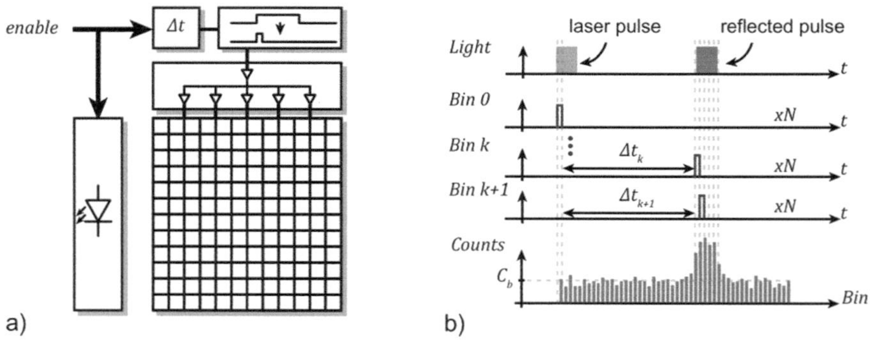

FIGURE 2.16 a) Simplified block diagram of the blocks involved in ToF computation in the SAER vision sensor. b) ToF computation principle.

2.3.5 Time-of-Flight calculation

Among the various techniques for ToF calculation, the SAER vision sensor employs a time-gating approach due to their advantages in terms of lower requirements at the pixel level, at the cost of requiring a larger number of laser pulses to achieve the same resolution than TDC approaches. TDC-based methods tend to be power-hungry and susceptible to background illumination, and their implementation at the pixel level often results in significant area consumption. In contrast, time-gating technique minimizes the additional resources needed at the pixel level, allowing for efficient ToF calculations while mitigating power consumption and area utilization challenges associated with other methods. By employing a resource-efficient approach, the SAER sensor offers the option for ToF calculations and keeps its versatility to work in different operation modes that can be suitable for different scenarios or applications.

As explained in Section 1.2.3, dToF can be measured by defining different time gates shifted with respect to the laser pulse. In this configuration, the *enable* signal serves as the input to the gating circuitry, which includes a programmable delay and a windowing circuit, as shown in Figure 2.16 (a). When the external control signal that triggers the laser is enabled, the gating circuitry generates a delay and controls the width of the gate that is applied to the pixels, as illustrated in Figure 2.16 (b). This is performed multiple times each time bin, creating a histogram that represents the convolution between the time gate and the light profile. By employing this approach, the SAER sensor effectively captures the temporal information required for precise ToF.

However, full histogram building is required as no mechanisms like zooming or sliding were implemented in this prototype. Therefore, there is still room for improvement in terms of ToF measurements for the SAER vision sensor.

2.3.6 Advantages of the SAER sensor in LiDAR systems

The asynchronous architecture of the SAER sensor offers several advantages in LiDAR systems, exploiting its asynchronous architecture and adaptive pixel behavior. By eliminating the need for reconstructing intensity maps in a frame format, the SAER sensor enables direct extraction of scene information, without continuous readout of meaningless pixels. This results in improved data efficiency and optimized memory requirements, crucial factors in LiDAR applications.

One advantage is its asynchronous acquisition of information, which aligns well with the nature of SPAD detectors and light. As photons arrive asynchronously, the SAER sensor can directly extract relevant information from the scene without the need for intensity map reconstruction. This asynchronous approach not only ensures efficient information acquisition but also enables online adjustment of pixel sensitivity. Also, the SAER sensor allows for the selective disabling of pixels, enabling efficient ToF calculations in specific *Regions of Interest* (ROI) or targeted pixels, further optimizing power consumption and enabling energy-efficient LiDAR applications. Indeed, a potential scenario could involve detecting intensity variations in DVS mode and selectively activating pixels in ToF mode that detected variations. By performing ToF calculations only when variations in the scene are detected, power consumption is significantly reduced, as the laser, the primary source of power consumption, is not enabled redundantly.

Another key aspect related to the asynchronous operation of the SAER vision sensor in the ToF is that pixels are asynchronously discharged. This means that pixels that did not receive any photons during a certain exposure are not scanned, avoiding expending energy and time in the process.

Furthermore, the online sensitivity adjustment capabilities provide additional advantages. With the ability to adapt the in-pixel thresholds, the sensor can implement advanced algorithms, achieve high dynamic range imaging, and adapt to varying ambient conditions. By dynamically adjusting pixel sensitivity, the sensor can extract meaningful information from the scene, enhancing its ability to capture relevant events and reduce unnecessary data transmission. This adaptability ensures optimal utilization of resources and improves the overall performance of the LiDAR system.

Finally, the implementation of a DVS allows the implementation of a motion-triggered ToF operation, aiming to optimize energy consumption of the system. Avoiding redundant laser exposures when the scene did not vary reduces considerably the power consumption of the system in idle state, since lasers are usually the dominant contributor to power consumption in LiDAR systems.

In summary, the SAER vision sensor offers significant advantages in LiDAR systems. Its asynchronous architecture, adaptive pixel behavior, and online sensitivity adjustment capabilities enable efficient information extraction, optimized power consumption, and improved data efficiency. By leveraging these advantages, the SAER sensor contributes to the development of energy-efficient LiDAR applications while enhancing the performance and adaptability of the system.

2.4 POWER MANAGEMENT UNIT

The PMU is a vital module in the SAER Vision system that is responsible for generating and regulating all the necessary power supply voltages from a single 5 V input.

In addition to the regular 3.3 V and 1.2 V that the system requires to work, the PMU also generates the required voltage to bias SPAD devices slightly above the breakdown voltage.

One fundamental aspect of SPAD-based imagers is this need for a high voltage power supply, usually in the range of tens of volts. This high voltage bias is crucial for proper operation and efficient photon detection. The PMU is in charge of regulating this voltage, which can be adapted by the CPM to adjust the excess voltage. This adjustability is a highly desired characteristic in SPAD-based systems since it is essential for fine-tuning the performance of the SPAD devices. By adjusting the voltage levels, parameters such as the PDP and DCR can be optimized to meet specific application requirements. This adjustability empowers system designers to achieve the desired balance between performance and power consumption, enhancing the overall efficiency of the SPAD-based system.

Another critical factor in SPAD-based systems is the variation of these parameters with temperature. The breakdown voltage of SPAD devices can change with temperature fluctuations, impacting their performance. Also, carrier thermal generation increases with temperature, increasing DCR [Xu17]. The PMU addresses this challenge by including this programmability in the SPAD voltage, since the CPM can use the temperature information provided from the SAER vision sensor to adapt this voltage accordingly. This temperature-based voltage adjustment helps maintain a consistent excess voltage, compensating for temperature-induced variations and ensuring stable and reliable SPAD device operation.

2.4.1 PMU Block diagram

The PMU Block diagram, as shown in Figure 2.17 illustrates the key components responsible for generating and regulating the power supply voltages in the SAER Vision system. The system features a 5 V input, which serves as the primary power source. This input is connected to a 3.3 V LDO, a 1.2 V LDO, and a DC-DC Boost converter.

The 3.3 V and 1.2 V LDOs receive the 5 V power supply as an input and provide stable and regulated voltage outputs. These voltages are utilized to supply the SAER vision sensor and additional circuitry present on the PCB, ensuring proper operation of the system.

Meanwhile, the DC-DC Boost converter is responsible for generating a higher voltage level required for the SPAD devices. The boost converter efficiently steps up the voltage from the 5 V input to 22 V. To reduce the noise introduced by the switched converter, an adjustable LDO is employed to regulate and stabilize the high voltage output. This noise reduction is critical for maintaining optimal performance of the power supply and minimizing any undesirable impact on the SPAD devices such an increase in the effective DCR induced by noise in the power supply.

Then, the output of the DC-DC converter is connected to a *High Voltage LDO* (HVLDO) that provides the required bias voltage for the SPAD devices. This regulated high voltage supply ensures that the SPAD devices are biased above their breakdown voltage and can be controlled by the CPM. Note that both the DC-DC converter and

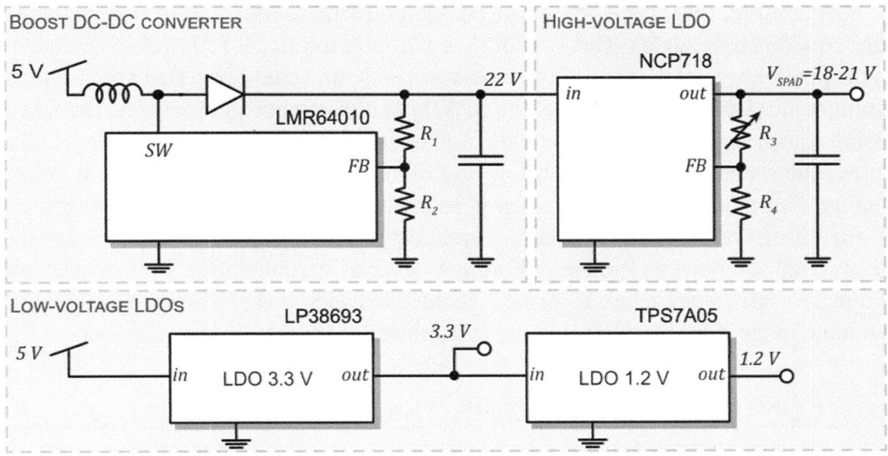

FIGURE 2.17 Block diagram of the Power Management Unit.

the HVLDO can be disabled to save power when the sensor is turned off. This power-saving capability enhances the overall energy efficiency of the system by ensuring that power is only consumed when necessary, helping to optimize battery life and reduce overall power consumption.

2.4.2 Adjustment of SPAD bias voltage

The boost DC-DC converter is a switched-inductor converter, where both the power MOSFET and the pulse-width-modulation circuit are located within the same part. The voltage at the output of the DC-DC converter is determined by the ratio of R_1 and R_2, which are selected to generate a voltage of 22 V. Similarly, the HVLDO determines the output voltage through the ratio of R_3 and R_4. However, in this case, R_3 is a programmable resistor that allows for the adjustment of the SPAD bias voltage by the CPM. Consequently, the output voltage of the HVLDO, V_{SPAD}, can be adjusted as follows:

$$V_{SPAD} \simeq 1.2\left(1+\frac{R_3}{R_4}\right) \tag{2.17}$$

The interaction between the CPM module and the programmable resistor provides a means to adapt the SPAD bias voltage based on real-time measurements and environmental conditions. For instance, the system can measure the temperature and dynamically adjust the SPAD bias voltage accordingly. This temperature-based adjustment helps to maintain the SPAD parameters, such as PDP and DCR, as constant as possible, ensuring consistent and reliable performance across varying operating conditions.

Furthermore, this configuration offers the potential to bias the SPADs as regular linear-mode photodiodes. This capability opens possibilities for future advancements in the pixel design, where the photodiodes can be optimized to operate in the accumulation mode [Ouh20]. By utilizing the SPADs as linear-mode photodiodes, the SAER vision sensor can extend its dynamic range under high-illumination conditions. This allows the sensor to capture both low and high levels of incident light with greater fidelity. This expanded dynamic range is valuable in a wide range of applications. It is worth noting that this may also require architectural modifications in the PMU design to account for power efficiency. Future work can explore these possibilities and refine the PMU architecture to strike a balance between power savings and extended dynamic range, further enhancing the capabilities of the SAER vision sensor.

2.5 CONTROL AND PROCESSING MODULE

The CPM plays a crucial role in the SAER Vision system, serving as the central control unit. Implemented on a FPGA board (Opal Kelly XEM7310) operating at 100 MHz, the CPM orchestrates the generation of control signals for various components, including the SAER vision sensor, the PMU, and the laser driver. Additionally, it handles data collection and processing from the SAER vision sensor, performing pre-processing tasks to extract meaningful information. The CPM acts as the bridge between the sensor and the USB host, facilitating the efficient transfer of processed data for further analysis or storage.

A highly efficient approach for the SAER Vision system would involve the utilization of a dedicated neuromorphic processor to handle the readout of spiking data from the sensor and perform subsequent processing tasks, including application-specific event generation (e.g., events related to object or pattern recognition) and sensitivity adaptation for each pixel. This neuromorphic processor would be specifically optimized for the specific requirements of the system, enabling real-time processing and advanced vision tasks. However, it is important to note that at the current stage of development, the system relies on an FPGA-based emulation of the neuromorphic processor interface, lacking the full asynchronous nature and neuronal network implementation. While these ideas provide promising directions for future research, the FPGA emulation serves as a practical solution to mimic certain aspects of the envisioned neuromorphic processor interface.

2.5.1 CPM BLOCK DIAGRAM

Figure 2.18 showcases the different blocks comprising the system inside the FPGA of the CPM. Firstly, the host interface enables communication with the host system through a USB 3.0 interface, facilitating data transfer and interaction between the SAER vision sensor and the software application. The SPI interface plays a crucial role in programming and configuring the sensor. It provides a convenient means of adjusting sensor parameters, such as sensitivity thresholds and bias configurations, offering flexibility and adaptability to different operating conditions.

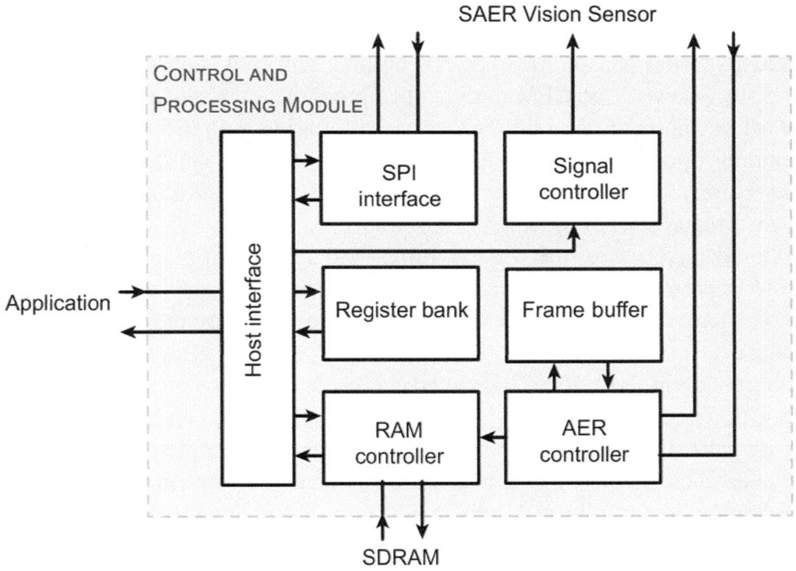

FIGURE 2.18 Block diagram of the control and processing module within the SAER vision system.

The register bank serves as a storage unit for vital information required during sensor operation. It holds data related to timing information for control signals and other global parameters, allowing for precise control and synchronization of the functionality of the SAER vision sensor. The signal controller is responsible for generating the necessary signals required by the sensor and laser driver to operate in different modes. It ensures proper timing and synchronization of signals.

The AER controller is a critical component that generates the signals required for the handshaking of the AER protocol. This enables the detection of events and their respective timestamping, providing valuable temporal information about the scene. Note that the AER controller can directly store these timestamps in the external memory to send raw data to the software application. However, the FPGA may also pre-process the data for an efficient operation before sending it to the application host.

Lastly, the *Frame Buffer* (FB) serves as a temporary storage location for frames reconstructed from incoming events. Depending on the operating mode, the information is pre-processed within the AER controller and then written to the FB. For example, in FR mode, the number of events is integrated into the FB, storing information related to frequency. In TNP mode, each pixel spikes once and its timestamp is stored in the FB. In discrete DVS mode it stores information regarding the reference inter-arrival time. In QI mode, the FB is utilized to integrate 1-bit images, while in ToF calculations, it helps to build the histogram bin for each pixel. The FB provides a convenient means of temporarily storing and transmitting pre-processed data to the software application for representation and analysis purposes.

2.5.2 On-line Sensitivity Adjustment

The SAER vision sensor incorporates on-line sensitivity adjustment capabilities [Gome23d], allowing the CPM to dynamically regulate the sensitivity of each pixel. Pixels are designed to program their sensitivity during the readout of events. In the free-running operation mode, pixels generate continuous spikes at a specific frequency. However, the CPM aims to equalize the bandwidth consumption of each pixel, ensuring efficient utilization of resources.

In the adaptive mode, the address corresponding to each pixel stores the photon threshold (represented by 3 bits) and the timestamps of the previous event, $t_{stamp}(k-1)$ (20 bits, for a dynamic range up to 120 dB). When a new event is received, the AER controller detects the arrival time of the event, $t_{sta}(k)$, and calculates the inter-arrival time of the current event, $T(k)$. If this time interval exceeds or falls below a threshold, th, with respect to the target event rate, T_{target}, the sensitivity is decreased or increased, respectively.

Figure 2.19 shows the block diagram of the algorithm implemented in the CPM, which takes 3 clock cycles, i.e., 30 ns. This dynamic adjustment mechanism allows for fine-tuning the sensitivity of individual pixels, optimizing their performance based on the characteristics of the observed scene. In practice, th must be defined taking into consideration the effect of photon shot noise and its relationship with N_{ph}. Thus, dark pixels must be compared to a higher threshold to avoid wrong decisions. This is accomplished by the factor β, which is dependent on the current value of N_{ph}. The condition the algorithm evaluates is:

$$|\Delta T| > th \tag{2.18}$$

where ΔT is the difference between T_{target} and $T(k)$. The value of th can be selected to be 3σ, where $\sigma = (k)/\sqrt{N_{ph}}$ and $(k) \simeq T_{target}$. Nevertheless, since these are constants, the value of β as a function of N_{ph} is taken from a look-up table.

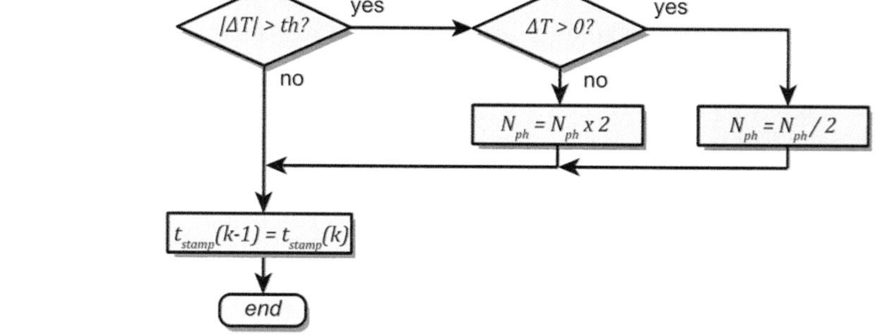

FIGURE 2.19 Algorithm implemented in the CPM for adaptive sensitivity adjustment.

Also note that the value of N_{ph} increases or decreases to the adjacent value. To achieve a faster settling, the best value of N_{ph} can be estimated from the target event rate and N_{ph}. However, this operation involves a division, which is not easily implementable in hardware.

Furthermore, the AER controller monitors the event rate of the readout channel. If the event rate approaches the saturation value, indicating a high volume of events, the controller sets N_{ph} to its maximum value, to minimize the time where the bandwidth of the readout channel is saturated. This approach ensures that the readout channel operates within its optimal range, minimizing the risk of saturation and maintaining reliable data transmission under sudden variations of the scene. It is also possible that even with the larger value of N_{ph} the readout channel saturates. In that scenario, pixels that are continuously readout can be temporarily disabled to allow the remaining pixels to transmit information. However, this complicates the operation, since reactivating the disabled pixels require a global reset operation or the execution of an SPI command, which interrupts the ongoing process.

Overall, the on-line sensitivity adjustment mechanism implemented in the CPM enables adaptive control of pixel sensitivity, ensuring efficient bandwidth utilization and robust performance of the SAER vision sensor in capturing and processing visual information.

2.5.3 Discrete Dynamic Vision mode

As detailed in Section 2.3.3, the SAER vision sensor classifies events into motion and intensity events, introducing the concept of a discrete DVS mode. In this mode, the CPM is responsible for extracting temporal details from these events and identifying changes in photon levels. The key component in this process is an event classifier, shown in Figure 2.20, which consists of a memory element and specialized processing components.

Within our discrete DVS framework, the primary signal of interest is the time required by a pixel to accumulate N_{ph} photons. Analogous to analog DVS systems,

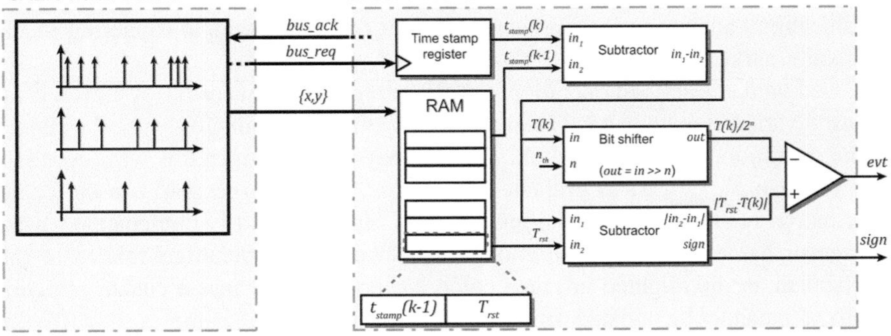

FIGURE 2.20 Block diagram of the event classifier implemented in the CPM for the discrete DVS mode.

TABLE 2.1
Bit description of the data stored in the discrete DVS mode.

Bits	Signal	Description
[19:0]	$t_{stamp}(k-1)$	Timestamp of the last event.
[39:20]	T_{rst}	Inter-arrival time of the last event detection.
[47:40]	n	Sample index.

where pixels reset after an event to store the current photoreceptor voltage, a digital version requires a memory element. This takes the shape of a digital register when working in the digital domain. In this implementation, we repurposed the FB to function as a per-pixel memory. Table 5.3 shows how the 48-bit element corresponding to each pixel address are distributed. It encompasses crucial parameters such as the inter-arrival time of the last motion event, T_{rst}, and the timestamp of the previous event, $t_{\text{stamp}}(k-1)$, used to compute the current inter-arrival time, $T(k)$. Additionally, a signal n is used as a counter to trigger the event classifier upon detection of a determined number of events, thus, virtually increasing the value of N_{ph} beyond the inherent limitations of the pixels.

The event classifier incorporates a subtractor to compute (k). Then, a two's complement subtractor computes the absolute value of the difference between (k). and T_{rst}, denoted as ΔT. Eventually, ΔT is compared to $\alpha_{\text{th}}(k)$ as demonstrated in equation (2.8). For simplicity, the value of α_{th} is implemented as a power of two. Thus, the bit-shifter is implemented by selectively neglecting the lowest significant bits of (k).

Figure 2.21 illustrates the operational flow of the event classifier. Upon receiving an event, it computes parameters (k) and ΔT. If $|\Delta T|$ exceeds $\alpha_{\text{th}}(k)$, the event is classified as a motion event and is relayed to the application. Simultaneously, $t_{\text{stamp}}(k-1)$ is updated and T_{rst} is set to $T(k)$.

This operation can be improved by including additional functionalities. Figure 2.22 shows an enhanced version of the algorithm. As mentioned earlier, N_{ph} can be virtually increased by using the information relative to more than one spike. To do so, the algorithm is triggered only when n equals a defined number of samples $N_{samples}$. Modifications relative to the implementation of this multi-sample triggering mechanism are marked in blue color.

Based on the described operation, the algorithm must wait for $N_{samples}$, even if the intensity varied enough to have been detected by the algorithm. To reduce the latency of the operation, a coarse threshold $\alpha' > \alpha_{\text{th}}$ may be defined. In this way, even if n is lower than $N_{samples}$, ΔT is evaluated. Therefore, a coarse threshold is used for the information related to fewer photons, while a fine threshold is implemented when the sensor has already gathered enough statistical data. Operations related to this mechanism are highlighted in green color. Nevertheless, in case a coarse detection occurs, T_{rst} cannot be updated, since it is not related to $N_{samples}$ samples. To solve this limitation, T_{rst} is set to 0 to force the initial state of the algorithm, represent as the red elements in Figure 2.22.

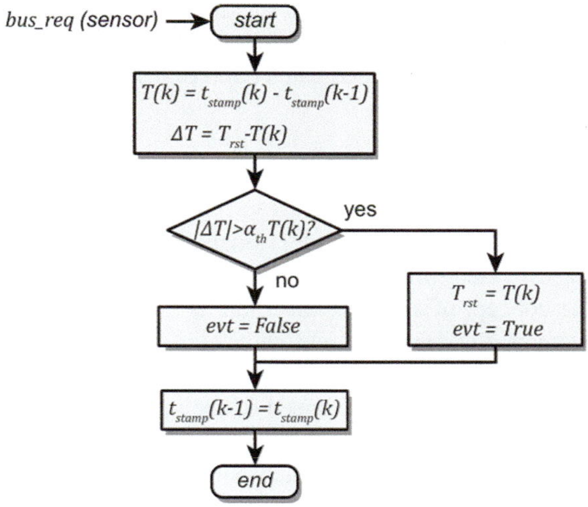

FIGURE 2.21 Algorithm flow chart implemented in the event classifier within the CPM.

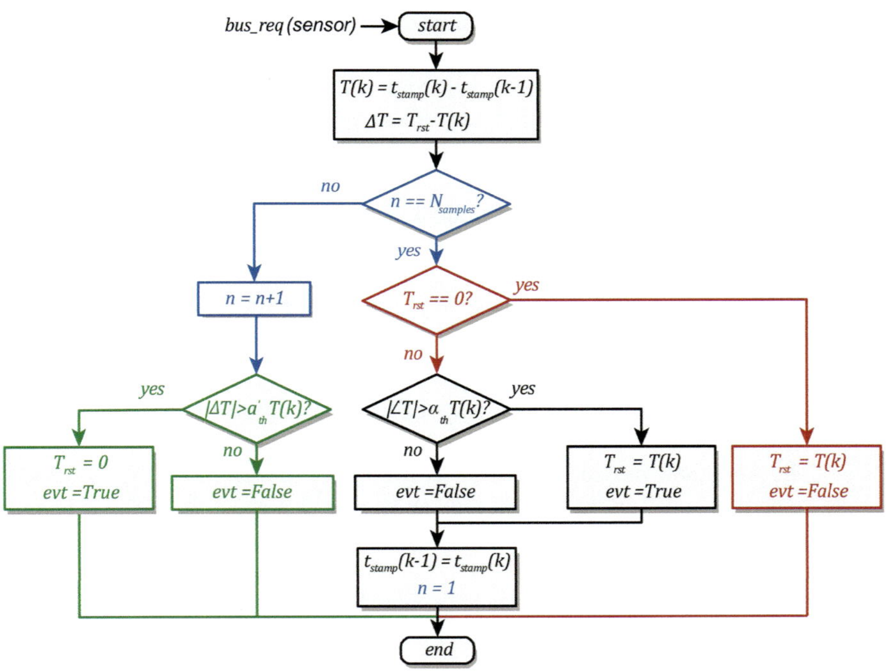

FIGURE 2.22 Flow chart of the extended algorithm including multiple-sample triggering and coarse detection.

Finally, note that if the event classifier was integrated within the pixel array or a lower die, the memory requirements per pixel would be notably reduced, as ΔT could be directly computed using a dedicated counter per pixel, reset after each event arrival.

2.6 SYSTEM PERFORMANCE

In the context of the SAER Vision System, evaluating its performance and comparing it to traditional SPAD-based cameras can be challenging due to the nature of its architecture, where the concept of a traditional frame is no longer applicable. However, to provide a meaningful comparison, we will define an equivalent frame rate that considers the necessary information required to reconstruct a frame.

One key performance metric is the speed of the system, which is determined by the frequency at which events are generated and processed. The SAER Vision System excels in this aspect by leveraging the asynchronous event-based approach, enabling high-speed processing and event detection with low latency.

Additionally, DR is a parameter of interest of a camera system since it provides information regarding its capability to detect bright and dark details in the scene. The time-encoding nature of the signals of the SAER Vision System inherently provides a high DR.

Another important metric is power consumption, which plays a crucial role in many applications. The SAER Vision System incorporates power-efficient design principles and optimization strategies, allowing for energy-conscious operation. By selectively enabling and disabling pixels, adjusting sensitivity, and adapting power supply voltages, the system achieves efficient power management.

This section outlines the limitations of these metrics at the system level, explaining their advantages compared to conventional frame-based camera systems. It also provides theoretical support for the system characterization, which is performed in Chapter 5.

2.6.1 Dynamic range

Two main factors limit the dynamic range of SPAD-based sensors:

1) The minimum measurable signal is limited by the DCR of SPAD devices, since it is added to the photon rate, appearing as a noise floor. Device engineering can be done to optimize this metric, achieving rates of below 1 $Hz/\mu m^2$ [Vorn21].
2) The upper limit of the dynamic range may be limited by different factors, such as the bit depth of the counter circuit or the dead time of the SPAD.

In conventional frame-based sensors, intensity levels are measured by counting photons in a pixel counter during an exposure time t_{exp}. As a result, the maximum photon count is limited by the bit depth of the counter, N, as shown in Figure 2.23 (a). Therefore, the number of bits of the counter must be increased to achieve a high dynamic range, with the consequent increase in area consumption.

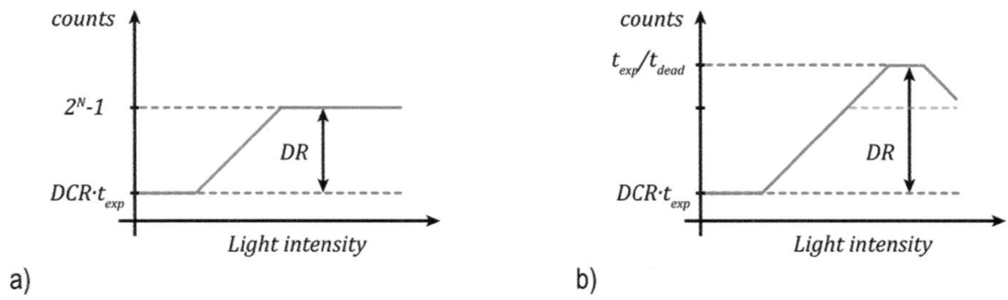

FIGURE 2.23 Relationship between light intensity and photon counts in conventional SPAD-based sensors: a) Counter limited. b) SPAD limited.

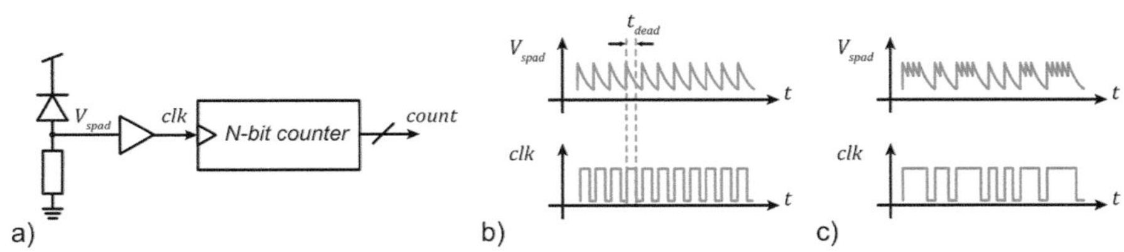

FIGURE 2.24 a) Photon counting using passively quenched SPAD. b) Waveforms at the highest measurable light intensity. c) Waveforms under saturation of the SPAD.

However, even if the counter does not limit the dynamic range, dynamics of SPAD impose an upper limit for the dynamic range. Figure 2.24 (a) shows a passively quenched SPAD device counting photons in an N-bit counter. Resulting waveforms under high illumination are displayed in Figure 2.24 (b). In this scenario, the inter-arrival time of photons is slightly larger than the dead time of the SPAD, t_{dead}. Thus, the maximum photon counts under these conditions are limited by t_{exp}/t_{dead} photons. If the intensity keeps increasing, photons arriving during t_{dead} are filtered out, as shown in Figure 2.24 (c). Therefore, the relationship between intensity and photon count is similar to that represented in Figure 2.23 (b).

Unlike conventional SPAD-based sensors, the SAER vision system does not read out the photon counts of an in-pixel counter. Indeed, there is no such concept as exposure time. On the contrary, the pixel output is a continuous pulse train that does not saturate depending on the bit depth of the in-pixel counter. This is attributed to the self-reset mechanism that occurs after event readout. Thus, the output of the SAER sensor shows the same relationship as in Figure 2.23 (b) but using the frequency (counts/s) as the magnitude of interest. The main advantage with respect to conventional architectures is that this relationship is achieved even using low values for N_{ph}, reducing the footprint area related to the counter.

2.6.2 Speed and Equivalent Frame Rate

Conventional frame-based sensors traditionally quantify their operational speed in terms of frame rate, given that data is transmitted as a complete image constituted by the entire pixel array. This frame rate, however, can be limited by various factors including pixel access time, A/D conversion, and the speed of input/output operations [Meyn09]. Moreover, rolling-shutter image sensors encounter limitations due to multiple exposure times, which inevitably decreases frame rate.

Unlike frame-based counterparts, where the bandwidth of the readout channel is uniformly distributed across the pixel array, asynchronous image sensors allocate this bandwidth in a more efficient manner. In the context of the SAER vision sensor, pixels detecting photons at a higher rate consume a larger portion of the available bandwidth. This dynamic allocation of bandwidth ensures a more adaptive approach to capturing dynamic visual information, setting it apart from traditional frame-based sensors.

Therefore, to evaluate the speed performance of the SAER vision system and determine an equivalent frame rate, it is important to define the smallest piece of information gathered from the sensor. Nevertheless, this definition is not trivial, and it is difficult to provide a general answer that applies to all possible scenarios or applications. In fact, the SAER vision sensor is conceived to operate in a continuous and asynchronous manner. Therefore, latency might be more important than the equivalent frame rate from an application point of view, since the equivalent frame is dynamically updated depending on the spiking data provided from the pixel array.

Denoting BW_{ch} as the bandwidth of the readout channel, the condition that the system must satisfy results:

$$\sum_{y}^{N}\sum_{x}^{M} f(x.y) < BW_{ch} \tag{2.19}$$

where N, M, and $f(x, y)$ are the number of rows, columns, and the spiking frequency of pixel (x, y) respectively. Thus, defining the equivalent frame rate as the time required to acquire information from all pixels, its maximum is achieved when BW_{ch} is evenly distributed across the pixel array. Also, the smallest piece of information corresponds to receiving two spikes from a pixel. In this way, the maximum achievable frame rate, fps_{max} is approximately:

$$fps_{max} \simeq \frac{BW_{ch}}{2NM} \tag{2.20}$$

However, this scenario may lack significance. If pixels exhibited identical sensitivity, the sensor would yield a plain image, which does not provide useful information. On the contrary, if pixels are configured with different sensitivities, details can be extracted even when all pixels spike at the same rate. In this case, dynamic range would be limited by the programmability range of the sensitivity.

At this point, the analysis becomes intricate since the acquisition speed is strongly dependent on the scene. If the information from all pixels is important and required for the application, the frame rate is limited by the spiking frequency of the darkest pixel.

Additionally, as the frequency of pixels decreases, it allows brighter pixels to spike at a higher rate. Hence, a correlation between dynamic range and acquisition time becomes apparent. To provide insights about the capabilities of the sensor, a basic analysis may assume, for simplicity, that half of the pixels operate under low illumination, while the remaining half function under high illumination. This scenario is equivalent to a linear gradient across rows and exhibits a maximum dynamic range is given by f_{max}/f_{min} where f_{min} equals the equivalent frame rate. Thus, f_{max} and the maximum dynamic range, DR_{max} can be expressed as:

$$f_{max} = \frac{BW_{ch}}{2NM} - f_{min} \tag{2.21}$$

$$DR_{max} \simeq K_s \left(\frac{BW_{ch}}{f_{min}} \frac{1}{2NM} - 1 \right) \tag{2.22}$$

Equation (2.22) suggests that the number of elements in the pixel array is also a variable that must be considered when comparing the speed of an asynchronous sensor to conventional frame-based architectures. Furthermore, since the SAER vision sensor allows adjusting sensitivity of each pixel individually, DR is extended by a factor K_s corresponding to the sensitivity ratio between bright and dark pixels.

In summary, the maximum equivalent frame rate and dynamic range are strongly correlated in spiking sensors, and dependent on the scene. Therefore, defining an equivalent frame rate is challenging. However, upper limits on this metric and its relationship with DR can be defined for the sake of comparison.

2.6.3 Power consumption

The SAER vision system encompasses several components that contribute to power consumption. Nevertheless, these contributions are somewhat correlated with each other, thus requiring specifying the operating conditions of the system when studying its energy consumption.

The greatest contribution to power consumption is associated with the CPM. This is because the CPM is implemented in a general-purpose development board. Also, the design within the FPGA is not optimized for power consumption and does not implement any power-saving mechanisms. Nevertheless, the CPM is expected to be substituted by a neuromorphic processor optimized to process spiking data in the most efficient manner. Therefore, the power consumption of the CPM is not considered in this analysis as it is outside of the scope of this work.

The next contributor is the optical emitter, since ToF operation requires multiple laser exposures. However, neither the peak power of the laser, P_{peak}, nor the number of exposures, N, is fixed, as it strongly depends on the target range, reflectivity, and desired accuracy and speed. Thus, the average power consumption of the laser, P_{laser}, can be estimated as:

$$P_{laser} = \frac{P_{peak}}{n} \frac{T_{pulse}}{T_{meas}} N \tag{2.23}$$

where η is the efficiency of the laser and T_{meas} is the acquisition time. If a more accurate measurement is desired, the SNR of the signal and thus N must be increased. Also, T_{pulse} must be reduced as much as possible. The minimum width of the laser pulse is limited by the laser dynamics and the width of the time bins in the histogram, as the laser pulse must fall inside two different bins at least to compute the CoM of the histogram.

Regarding the SAER vision sensor, its power consumption is dependent on the event rate, consequently varying based on the scene being captured. This behavior is similar to that found in asynchronous image sensors found in the literature and can be associated with a varying switching rate of SPAD signals, internal nodes, and output pads. Thus, the power consumption is expected to increase linearly with the event rate. Power breakdown among different elements of the SAER vision sensor will be performed in Chapter 5.

Finally, the power consumption of the PMU must be associated with its efficiency. The PMU converts the input DC voltage to the suitable levels required by each component. Nevertheless, there are losses associated with the switched converter and dropout voltages of regulators, which are dependent on the load, and therefore, on the SAER vision sensor.

2.7 SUMMARY

The SAER Vision System presents an alternative approach in the field of LiDAR systems, showcasing innovative components and functionalities. This chapter provided an overview of the key features and advantages of the SAER Vision System, including the SAER vision sensor, optical emitter, PMU, and CPM.

The SAER vision sensor features a first-of-its-kind asynchronous SPAD-based architecture that enables direct extraction of photon information, eliminating the need for scanning all pixels to extract frames corresponding to intensity maps in conventional SPAD-based sensors. Conversely, it enables continuous operation that benefits from the advantages of time-encoded signals. Its adaptability and pixel sensitivity adjustment contribute to efficient power consumption and bandwidth utilization.

Furthermore, a dynamic approach for dynamic vision that fits with the concept of the SAER vision sensor has been proposed and modeled. This system represents the first camera that detects motion using SPADs in an event-driven manner. Leveraging this operation mode, an efficient utilization of the laser allows optimizing energy efficiency of the system, especially when the scene is static and all laser exposures for ToF measurement are redundant.

The PMU ensures reliable power supply voltage generation and regulation for all system components. Its adjustability enables precise control of the bias voltage for SPADS, enhancing performance under temperature variations, since compensating for variations in the SPAD breakdown voltage is possible.

The CPM, implemented on a FPGA board, generates control signals, collects, and processes data from the SAER vision sensor, and enables pre-processing for efficient information transmission to the host device. While the current implementation emulates a neuromorphic processor, it lays the groundwork for future integration of dedicated processors, optimizing data processing and adaptability.

In summary, the SAER Vision System offers efficient information acquisition, adaptive sensitivity control, optimized power consumption, and accurate depth measurements. Its unique features facilitate energy-efficient LiDAR applications, enhanced dynamic range imaging, and advanced vision tasks.

3 The SAER vision sensor

The key element of the proposed system is the 3D vision sensor, called the SAER vision sensor. This sensor contains a 64×64 pixel array that detects impinging photons down to the single-photon level. Measured data is pre-processed at the pixel level and compressed in the form of events to finally send the information to the CPM. Pixel events are conveyed using the well-known AER protocol, implementing a word-serial address-event link. The sensor can work both in 2D and 3D mode. In the former, the sensor outputs data related to the luminance level of the scene. The 2D information can be obtained using Photon Counting [Dutt16; Panc13] or QI [Foss13] techniques. In 3D mode, the sensor measures the ToF of the emitter using time-gating techniques [Gyon23; Mori20a; Rocc20]. Sensors implementing these techniques benefit from a reduced in-pixel circuitry and higher background-illumination rejection at the expense of increasing the latency of the measurement.

3.1 BLOCK DIAGRAM OF THE IC

The SAER vision sensor includes different blocks to carry out and support the operation of the SAER pixel. Figure 3.1 shows the block diagram of the vision sensor, namely:

- The 64×64 pixel array. Each pixel contains a 4-bit *Static Random Access Memory* (SRAM) for configuration purposes.
- The AER readout block generates the signals that control the readout of the pixel array, in addition to the signals that request access to the external receiver. It writes the coordinates of the pixels in the address bus.
- The SRAM controller generates the signals to write the in-pixel SRAMs.
- The programmable delay line shifts the external enable signal to implement time-gating techniques.
- The bias generator generates reference voltages and currents.
- The digital core includes a SPI driver and a register bank to program the chip and carry out basic operations.

DOI: 10.1201/9781003427490-3

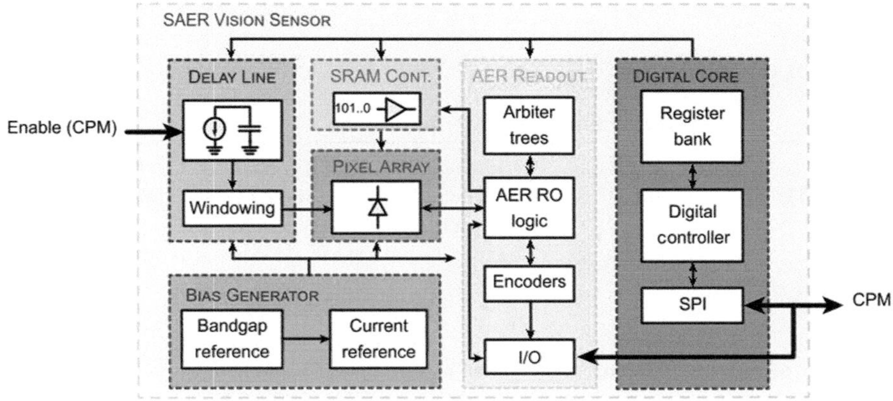

FIGURE 3.1 Block diagram of the SAER vision sensor.

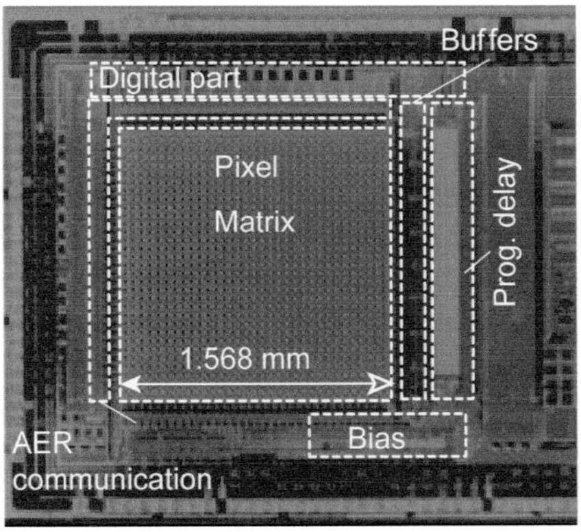

FIGURE 3.2 Microphotograph of the SAER vision sensor, highlighting its different blocks.

Figure 3.2 shows a microphotograph of the SAER vision sensor, highlighting the different blocks previously described.

3.2 PIXEL ARRAY

Pixels are the core of the SAER vision sensor. They are in charge of sensing the intensity of the scene by detecting individual photons. The pixel has two main interfaces: the AER readout logic for event transmission and the SRAM interface, to update configuration of individual pixels.

Table 1.1 summarizes the inputs and outputs of the pixel array. The operation of the pixel array is controlled by the global *enable* signal, which is the output of the programmable delay. The in-pixel SRAM allows disabling user-defined pixels and defining a ROI in the pixel array, in addition to programming different functionalities

TABLE 3.1
Signals of the pixel array.

Signal	Direction	Description
enable	Input	Enable signal for the pixel array. Global.
wr_col[31: 0]	Input	Write enable signal for the in-pixel SRAM. Shared every two columns. Active low.
d0[31: 0]	Input	Input data of the SRAM. Shared every two rows.
d1[31: 0]		
d2[31: 0]		
d3[31: 0]		
req_row[63: 0]	Output	Row request lines. Shared by row.
req_col[63: 0]	Output	Column request lines. Shared by column.
ack_row[63: 0]	Input	Row acknowledge signals. Shared by row.
rst_row[63: 0]	Input	Reset row signals. Shared by row.
rst_col[63: 0]	Input	Reset column signals. Shared by column.

of the pixel. The writing operation of the SRAM is controlled by the bus *wr_col*, where each element is shared every two columns. The input data signals, *d*0, *d*1, *d*2, and *d*3 are shared every two rows. Section 3.3.2 will delve deeper into their operation.

The 64-bit output buses, *req_row* and *req_col*, are used to implement the 4-way handshake of the AER protocol. They implement a wired OR at the row and column level, respectively. On the other hand, the input *ack_row* bus indicates which row is granted access to the readout channel using a one-hot encoding. Finally, *rst_row* and *rst_col* buses, both using one-hot code, reset the state of the pixel after its address is readout by the external receiver. The writing operation of the pixel (i, j) is performed when the signals *rst_r*[*i*], *rst_col*[*j*] and the corresponding bit of *write_sram* are activated at the same time.

3.3 AER READOUT AND SRAM CONTROLLER

The SAER vision sensor departs from conventional image sensors in that readout is not externally controlled, but pixels autonomously request access to the readout channel once they have gathered sufficient information. Moreover, the in-pixel memory undergoes updates concurrently with the readout process, requiring the seamless integration of the SRAM controller with the AER readout block. This section describes the circuitry within these blocks, as well as its operation.

3.3.1 AER READOUT BLOCK

The AER readout circuitry processes the output of the array and generates the signals mentioned in Section 3.2 that implement the handshaking. The AER readout block is identical at the column and at the row level and its operation has been demonstrated in previous event-based sensor implementations [Gome23a; Gome23c; Lene17]. Table 3.2 and Figure 3.3 summarize the signals and circuits involved on its operation, respectively. This block is composed of:

TABLE 3.2
AER readout block signals.

Signal	Direction	Description
***req*[63: 0]**	Input	Row/column request from the pixel array.
***ack*[63: 0]**	Input	Row/column acknowledge to the pixel array.
***rst*[63: 0]**	Input	Row/column reset signal to the pixel array.
req_aer	Output	Bus request at the row/column level.
bus_ack	Input	Bus acknowledge signal from the external receiver. Active low.
***addr*[5: 0]**	Output	Address bus with the coordinates of the acknowledged pixel.
wr_sram	Input	SRAM write enable signal.
rst_glob	Input	Global reset signal.
***rst_dig*[63: 0]**	Input	Row/column reset signal from the digital core.

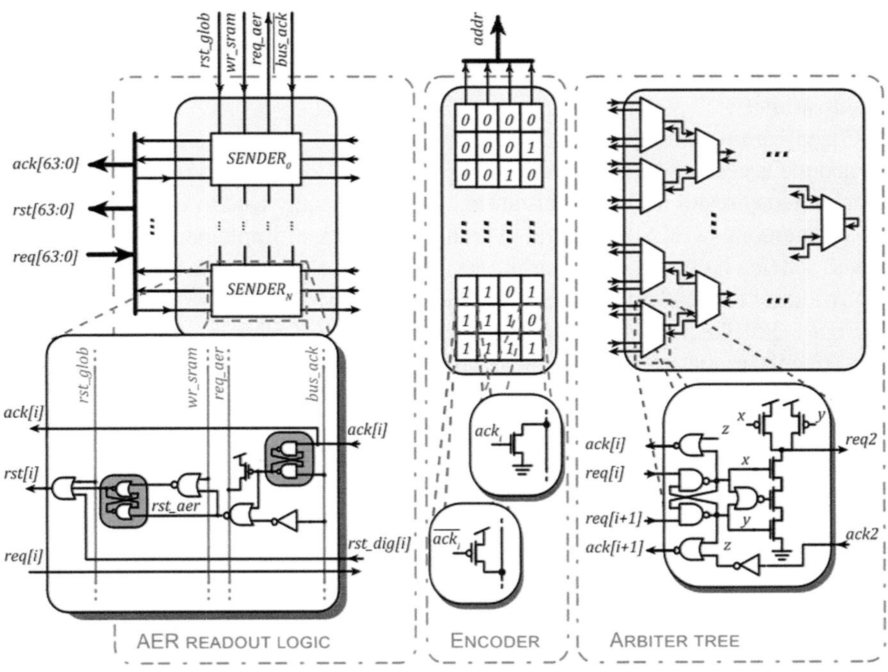

FIGURE 3.3 Block diagram of the AER readout block.

- A six-level greedy arbiter tree.
- An encoder that converts the one-hot code into binary. The outputs of the row and column encoders are the output of the sensor.
- 64 AER readout logic blocks, also called *senders*.

Figure 3.4 depicts the evolution of the signals involved in the AER readout block. The arbiter tree decides which row or column accesses the readout channel depending on the arrival time of the elements of the *req* bus (*req_row* and *req_col*). The output of this circuit, i.e., the *ack* bus (*ack_row* and *ack_col*) is the input of the encoder, which writes the coordinates of the pixel in the *addr* bus. Finally, the selection logic block generates the global signals *req_aer_y* and *req_aer_x* at the row and column level, respectively. The AND operation of these two signals generates the *bus_req* signal to request access to the external receiver. When the receiver acknowledges the request, i.e., activates the *bus_ack* signal, the selection logic block activates *rst_aer*, which is used to generate *rst* (*rst_row* and *rst_col*). This signal is latched to keep the pixel reset while updating the SRAM data. In this way, *rst* is only disabled when the global signal *wr_sram* is brought to a low state. This guarantees proper writing operation. Furthermore, *rst_row* and *rst_col* can also be triggered by a bus coming from the digital core (*rst_dig_row* and *rst_dig_col*) or the global *rst_global* signal. This allows random access to the pixel array and a global writing operation.

The choice of greedy arbiters is based on the higher bandwidth they offer. This is due to the fact that fair arbitration schemes prioritize equal access for all pixels, which involves additional checks and considerations to ensure fairness that compromise speed. Thus, they offer a slightly slower arbitration process compared to greedy arbiters.

Greedy arbiters, on the other hand, prioritize speed and efficiency, often granting immediate access to the pixel with the highest priority or earliest request. While this approach may result in faster arbitration, it can potentially lead to uneven access distribution among pixels. Therefore, the main drawback of implementation of an arbitration scheme based on greedy arbiters is that a few elements may occupy the entire bandwidth of the readout channel if their event rate is high enough.

Figure 3.5 illustrates the described scenario with a 4-element arbiter tree, where all elements request access to the readout channel simultaneously. Initially, one

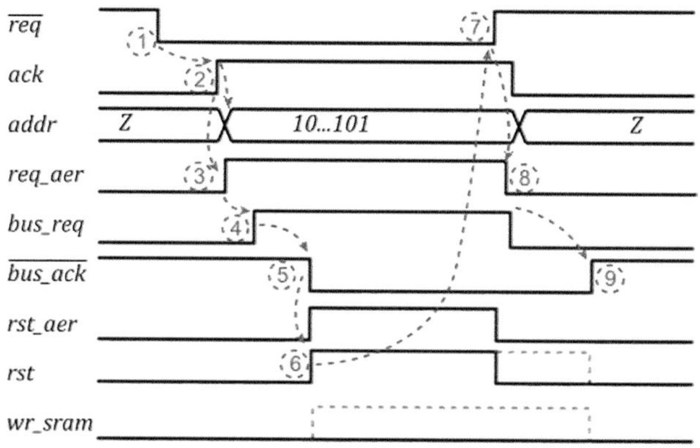

FIGURE 3.4 Waveform diagram illustrating signals in the AER readout block.

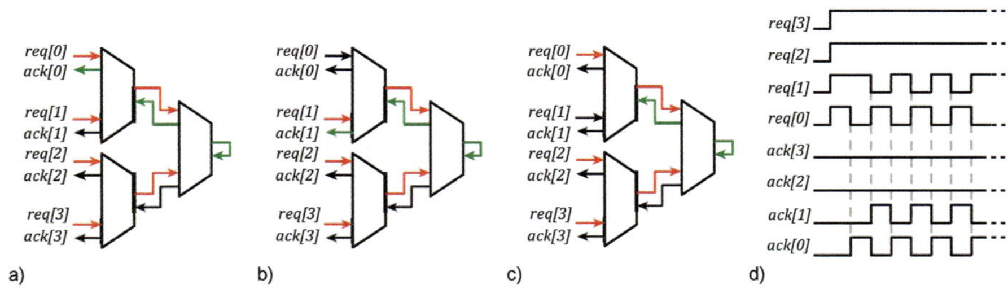

FIGURE 3.5 Bandwidth allocation in the readout channel (red: requested path, green: granted path). a) Simultaneous access request, Element 0 granted. b) Element 0 released, neighbor acknowledged. c) Element 1 releases, Element 0 re-requests. d) Resulting waveforms.

element is granted access (element 0 in Figure 3.5 (a)). Subsequently, the neighboring element gains access as most of the path is already granted (Figure 3.5 (b)). However, if element 0 requests access before the path is released (as in Figure 3.5 (c)), it takes priority over the other elements. This leads to a toggle between elements 0 and 1, resulting in a *pin-pong* effect, where information from the remaining elements is never read out (Figure 3.5 (d)). In the particular case of spiking pixels, this effect is observed when highly illuminated spots lead to a few rows or columns transmitting information, while the remainder waits indefinitely.

On the other hand, the advantage of greedy arbiters resides in this fast response when the path is partially granted. When two neighboring elements simultaneously request access, the time interval between both accesses is reduced to the delay of a single arbiter. This characteristic is particularly beneficial in image sensor applications, where adjacent pixels often generate events close in time if spatial resolution is sufficiently high.

3.3.2 SRAM CONTROLLER

Pixel memory in the SAER vision sensor can be updated with the value of $d3$-$d0$ after readout, i.e., when the pixel is reset. This means that pixel (i, j) is updated when $rst\_r[i]$ and $rst\_col[j]$ are activated, along with the corresponding bit of wr_col. Hence, the SRAM controller is responsible for generating these signals following readout when the user requests for memory update. Table 3.3 summarizes the signals involved in the operation of the SRAM controller.

The SRAM controller is shared every two rows and two columns. Figure 3.6 shows its implementation at the row level. It is composed of a 4-bit register and four drivers connected to $d0$-$d3$. These registers are level-sensitive D latches that are updated when wr_sram and any of the two rst_row bits are activated. The input can be connected to din_global to update the memories with global data under the control of the CPM, or to din_dig, which is controlled by the digital core and can be independently configured for every two rows. The multiplexer is controlled by the $cfg\_sram\_data\_global$ signal, which is stored in an internal register that can be updated via an SPI command.

TABLE 3.3
SRAM controller signals.

Signal	Direction	Description
***rst_row*[63: 0]**	Input/Output	Row reset signals to the pixel array.
***rst_col*[63: 0]**	Input/Output	Column reset signal to the pixel array.
wr_sram	Input	Row/column reset signal to the pixel array.
***din_global*[3: 0]**	Output	Global input data for pixel SRAM.
***din_dig*[31: 0][3: 0]**	Input	Input data for pixel SRAM from digital core. Shared every two rows.
***d0*[31: 0]** ***d1*[31: 0]** ***d2*[31: 0]** ***d3*[31: 0]**	Output	Input data of the SRAM. Shared every two rows.
***wr_col*[31: 0]**	Output	SRAM write flag.
cfg_sram_data_glob	***a*** Input	Configuration bit to select the input of the SRAM controller. 0: Internal data (from digital core). 1: Global input data (from CPM).

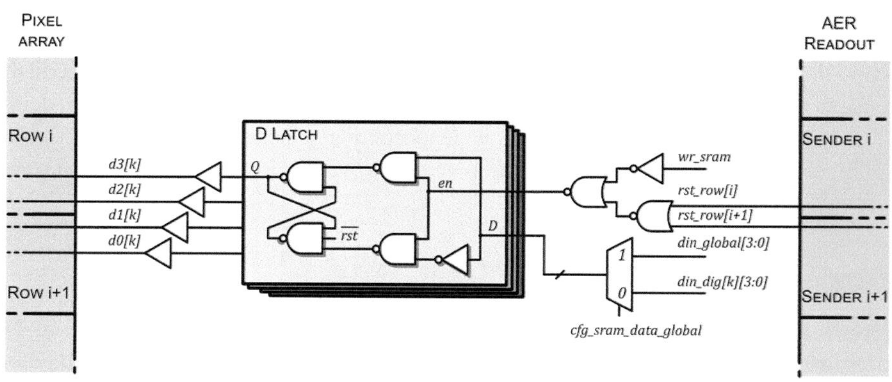

FIGURE 3.6 Schematics depicting the SRAM controller shared between every two rows.

At the column level, the SRAM controller generates *wr_col* when *wr_sram* and any of the two bits of *rst_col* is activated, as shown in Figure 3.7 (a). However, if the user initiates a global writing operation, it can result in a substantial current draw in the pixel array. This may cause voltage drops at the power supply, potentially leading to incorrect voltage levels at the SRAM. Such deviations can result in erroneous writing operations or even metastability, characterized by a large current flow through the pixel array. This will be further developed in Chapter 4.

To mitigate this, the column sender in the AER readout block is slightly modified, as illustrated in Figure 3.7 (b). An internal delay of approximately 1.5 ns implemented with an inverter chain is introduced between columns, ensuring that

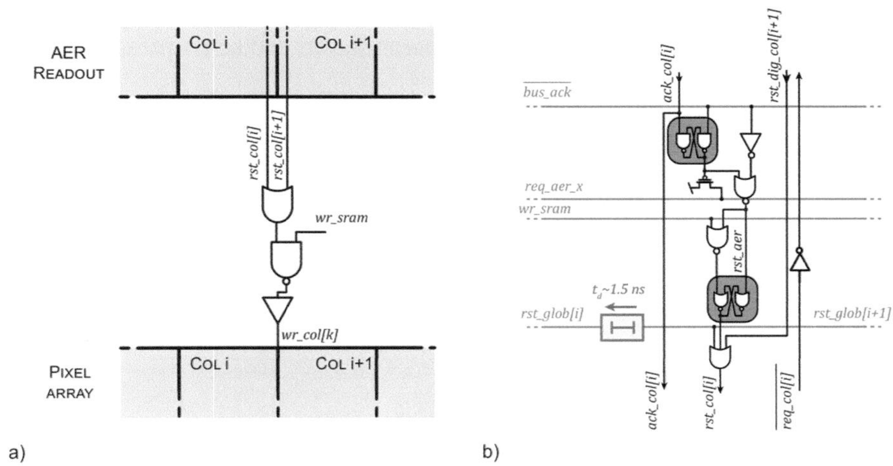

FIGURE 3.7 a) Schematics depicting the SRAM controller shared between every two columns. b) Column sender including delay for global reset.

$rst\_c[i]$ is activated before $rst\_col[i + 1]$. This staggered writing operation guarantees that the SRAM of each column has settled before starting the write operation in the subsequent column.

3.4 LOAD AND TIME DELAY REDUCTION IN AER SENSORS

The SAER vision sensor relies on an AER interface that implements a four-phase handshaking protocol [Boah00], detailed in Section 3.3.1. Within this protocol, once a pixel generates an event, the sensor requests access to the readout channel. Upon acknowledgement from the receiver, row and column addresses corresponding to a pixel generating an event are transmitted in parallel.

Each element of an asynchronous sensor encodes its information in the form of events. Then, this information is sent outside the sensor using a common readout channel, which access is controlled by an arbitration circuit. However, scaling up the array size presents limitations, particularly related to the impedance of the common lines at the row and column level. With increasing array size and technology shrinking, these impedances might constrain the transmission of events to the periphery, subsequently reducing the channel bandwidth.

Event information is encoded in the time or frequency domain and sent through the shared readout channel. The receiver needs to process this data in real time or store the timestamp of each event to prevent data corruption. Therefore, the primary bottleneck emerges in the bandwidth of the receiver. Assuming the receiver can handle all incoming events, the next limitation arises in the arbitration circuit. This circuit manages access to the readout channel when multiple elements request it simultaneously. However, in instances of numerous simultaneous access requests, collisions in the readout channel can cause temporal information loss due to arbitration time. Despite increasing the number of arbitration stages when doubling the element count, the arbitration time does not linearly increase with stage numbers. This occurs because the length of wires between stages, and hence parasitic capacitance, also doubles.

Limitations also manifest at the pixel level, even when the arbitration time might be negligible for a specific event rate. Pixels start the protocol requesting access to the periphery. These requests are initiated using per-row and per-column signals, implementing a wired NOR gate, requiring a single NMOS transistor per dimension within each pixel. Externally to the pixel array, a pull-up device (such as a PMOS transistor or resistor) keeps the voltage at a high level when no pixel seeks to be read out. Ideally, the request signal should promptly drop after an event occurs. However, in image sensors with numerous elements sharing the request signal, the impedance of this line becomes significant.

3.4.1 Timing limitations of the AER communication

Most temporal limitations that appear when increasing the number of elements in a system reliant on the AER protocol are related to the implementation of the wired NOR gate [Gome23e]. Figure 3.8 (a) depicts how this wired NOR gate is implemented across the pixel array. This scheme allows increasing the number of elements in one dimension without the need to include additional wires.

Apart from the pull-down and pull-up transistors $M_{pd,}$ and M_{pu}, respectively, the circuit can be modeled by including a unit resistance R_o and capacitance C_o per pixel. Also, the ground wire presents a unit resistance R_{gnd} that cannot be neglected when the number of elements is large.

The pull-down time constant, $\tau_{req,}$, is affected by R_o and C_o. The worst-case scenario occurs when an event occurs in the farthest pixel from the peripheral readout circuitry. Assuming that ground resistance is negligible, $\tau_{req,}$ can be approximated using the Elmore delay model [Elmo48]:

$$\tau_{req,pd} \simeq N\left(\frac{N+1}{2}R_o + R_{pd}\right)C_o \simeq \frac{N^2}{2}R_oC_o \tag{3.1}$$

where N is the number of elements in the row and column and R_{pd} is the on resistance of the pull-down transistor, which can be considered negligible if N is large enough. Note that if this occurs, the time constant increases as the square of N.

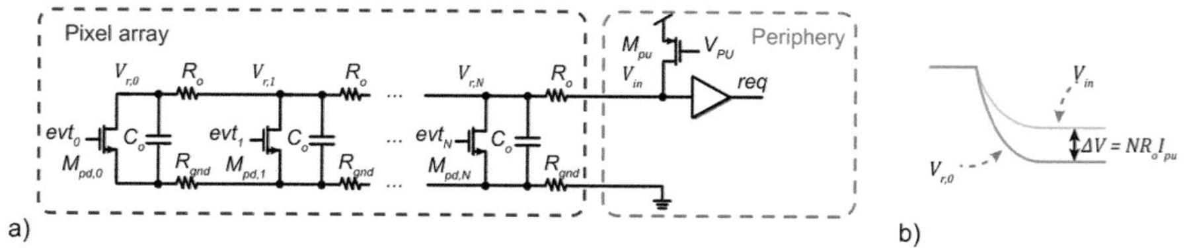

FIGURE 3.8 a) Wired NOR implementation across the pixel array. b) Voltage drop due to line resistance.

Not only the delay, but also the voltage level at the input of the arbitration circuit is an issue, as shown in Figure 3.8 (b). If the pull-up time is to be constant, the pull-up current must increase with N. Since the point-to-point resistance also increases with N, the logic-low voltage increases as N does. This is problematic in technology nodes where the supply voltage of the logic circuits is low, defining an upper limit for the pull-up current and, in consequence, limiting the bandwidth of the sensor. Besides, to set a well-defined logic-low voltage, R_{pd} must be sufficiently low. The optimum width of the pull-down transistor is that at which the drain capacitance begins to dominate over the routing capacitance since after that point the transistor becomes self-loaded, and there is no benefit when increasing its size. Hence, the driving capability of this transistor also limits the maximum pull-up current that can be used.

Problems also arise when R_{gnd} cannot be neglected. If the pixel array is connected to ground at both ends, the worst case is now in the middle pixel. In large arrays, the on resistance of the pull-down transistor increases due to the voltage drop across the ground resistance, leading to a longer delay and even making it impossible to reach the logic-low voltage level. Therefore, the pull-up current and the pull-up time are also limited by the ground resistance.

3.4.2 Proposed solutions

The limitations presented in the previous section are related to the impedance of the request line. Although this impedance is limited by the pitch of the pixels, routing congestion, and technology parameters, some design considerations can reduce its impact on the circuit performance. The following techniques are discussed in this section:

- Reducing the number of elements per readout channel.
- Reducing the effect of the line impedance.
- Splitting the pull-up element.
- Using adaptive pull-up transistors.

Detailing and analyzing these techniques will help determine their advantageous implementation scenarios compared to conventional methods.

Reducing the number of elements

The most obvious solution to improve the timing performance of the system consists of including multiple readout channels, each connected to a portion of the array.

For instance, the pixel array of the sensor from Figure 3.9 (a) implementing a single readout channel can be split into different subarrays, each with a different readout channel. Figure 3.9 (b) depicts a sensor where the pixel array is divided into four different subarrays, SA_i, each connected to its own AER readout circuitry. In fact, conventional image sensors regularly include two readout channels in parallel to increase the effective data rate [Funa15; Okad21; Tots16]. Indeed, 3D stacking technology has enabled image sensor architectures where ADC arrays are available [Saka18].

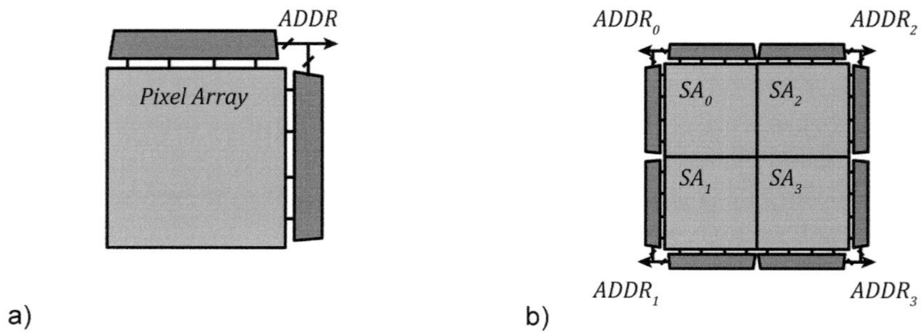

FIGURE 3.9 a) Sensor using a single readout channel for the entire pixel array. b) Sensor using four readout channels, one for each subarray.

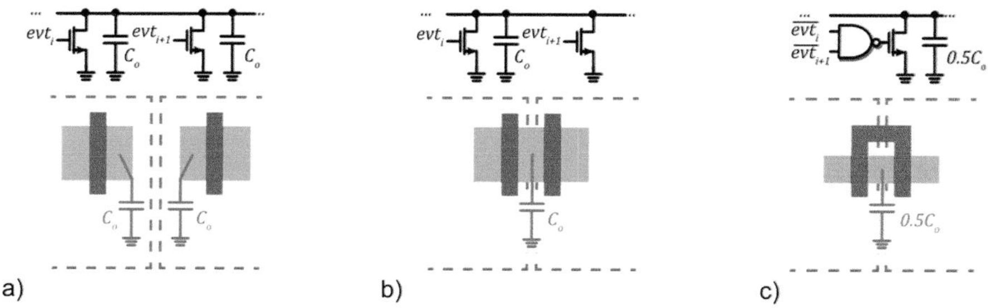

FIGURE 3.10 Alternative layout for the pull-down transistor. a) One transistor per pixel. b) Neighboring pixels sharing one diffusion. c) Neighboring pixels sharing pull-down transistor.

However, while conventional image sensors implement high-speed serial links to transmit pixel data [Funa15; Okad21; Saka18; Sato20; Tots16; Zhan19b], event addresses are transmitted in parallel. Thus, the use of multiple readout channels would considerably increase system complexity since additional pads and handshaking logic are required. To alleviate this drawback, each channel can be connected to a serializer [Miro07], which may also reduce the constraints associated with the external receiver.

Reducing the effect of the line impedance

To minimize $\tau_{req,}$ it is essential to decrease R_o and C_o. Reducing R_o can be achieved by increasing the trace width or decreasing the pixel pitch, but these adjustments might not be feasible in all scenarios. Regarding C_o, it has two main contributions: the capacitance associated to parallel traces and the drain capacitance of the pull-down transistor. The first contribution might not be reduced if distance among traces is limited by the pixel pitch; however, drain capacitance of pull-down transistors can be minimized.

This capacitance can be reduced by proper layout techniques, i.e., minimizing the area of diffusion. Figure 3.10 (a) shows the layout of the pull-down transistors when each uses a one-finger transistor. Capacitance can be reduced by sharing the drain diffusion, as in Figure 3.10 (b), which is equivalent as using a two-finger device in each pixel.

Further reduction can be achieved by sharing the pull-down transistor among adjacent pixels. Figure 3.10 (c) illustrates two neighboring pixels sharing the pull-down transistor, activated by the NAND operation of the two events. Although this modification increases the number of transistors, it effectively diminishes the contribution of the pull-down transistor in C_o by 50%, consequently reducing the current of the pull-up transistor, I_{pu} by the same amount. Therefore, resource sharing addresses all limitations discussed in Section 3.4.1.

Another efficient technique to diminish $\tau_{req,}$ involves inserting repeaters into the request line [Zhan19a]. Figure 3.11 illustrates the implementation of a repeater. This method approximates the time constant of each group τ_k, and the overall equivalent time constant, τ'_{req}, as follows:

$$\tau_k \simeq \frac{N}{k+1}\left(\frac{\frac{N}{k+1}}{2}R_o + R_{pd}\right)C_o \simeq \frac{\tau_{req,pd}}{k^2} \tag{3.2}$$

$$\tau_{req,pd} \simeq k\left(\tau_k + \tau_{inv}\right) \simeq \frac{\tau_{req,pd}}{k} + k\tau_{inv} \tag{3.3}$$

where k is the number of repeaters included within the row or column and τ_{inv} is the time constant of the inverter within the repeater. Equations (3.6) and (3.7) show that the pull-down time is reduced as k increases, until the propagation delay of inverters dominates. Thus, the optimal number of repeaters, k_{opt}, depends on τ_{inv} and the time constant of each pixel:

$$K_{opt} - \sqrt{\frac{T_{req,pd}}{T_{inv}}} - N\sqrt{\frac{R_o C_o}{2\tau_{inv}}} \tag{3.4}$$

The simulation results confirming the behavior of the pull-down transistors described in equations (3.1) and (3.3) are depicted in Figure 3.12. The values of R_o

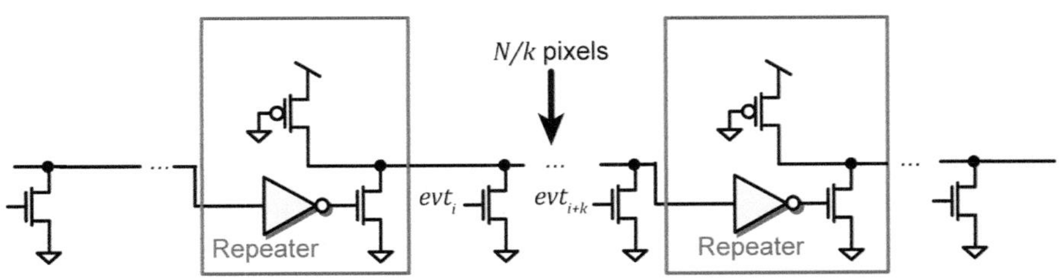

FIGURE 3.11 Insertion of repeaters inside the request line to reduce fanout.

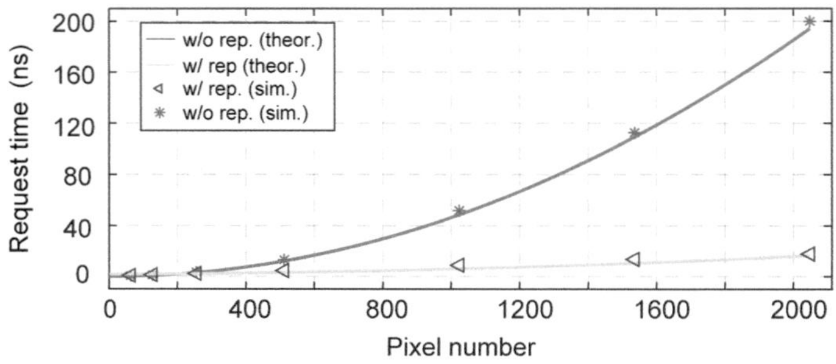

FIGURE 3.12 Simulated request pull-down including a repeater every 64 elements.

and C_o were extracted from the layout view of a pixel of the SAER vision sensor, and repeaters were inserted every 64 pixels. The fitting of the model to the simulation results is evident, demonstrating a significant enhancement in the timing performance of the request line with the inclusion of repeaters. Notably, for a number of elements lower than 400 pixels, the improvement remains marginal.

Splitting the pull-up transistor

As previously mentioned, the voltage drop across the non-zero resistance of the line can worsen the sensor performance. This is mainly because the pull-up transistor is usually located next to the arbitration circuitry for simplicity, as shown in Figure 3.8.

The pull-up transistor can be located in the opposite side to make the circuit insensitive to this voltage drop. However, the pull-up time increases due to a non-uniform current flowing through each parasitic capacitance.

To further enhance the performance of the circuit, the pull-up transistor can be split into both sides of the array. Thus, the resistance seen from the pull-up transistor is reduced by a factor of four, and the current flowing through the request line is halved. The pull-up transistor can also be split inside the pixel array to further reduce the current flowing through the entire metal trace; however, the area of the pixel increases.

Using adaptive pull-up transistors

The power consumption associated with the common pull-up networks of AER sensors increases linearly with the number of rows and columns when the pull-up current is constant. Nevertheless, as the number of pixels increases, so does the line capacitance, and thus the delay increases when using a constant-current approach. Although linearly increasing the pull-up current keeps the pull-up time constant, it might not be possible, as discussed in Section 3.4.1. This approach is also inefficient, as DC current is wasted when the signal is in the pull-down state. Ideally, current should only flow when all pull-down transistors are OFF. To overcome these limitations, an adaptive pull-up network can be implemented.

The proposed pull-up element consists of two pull-up transistors; M_w is a weak transistor that keeps the signal at a logic-high voltage while there is no request, and

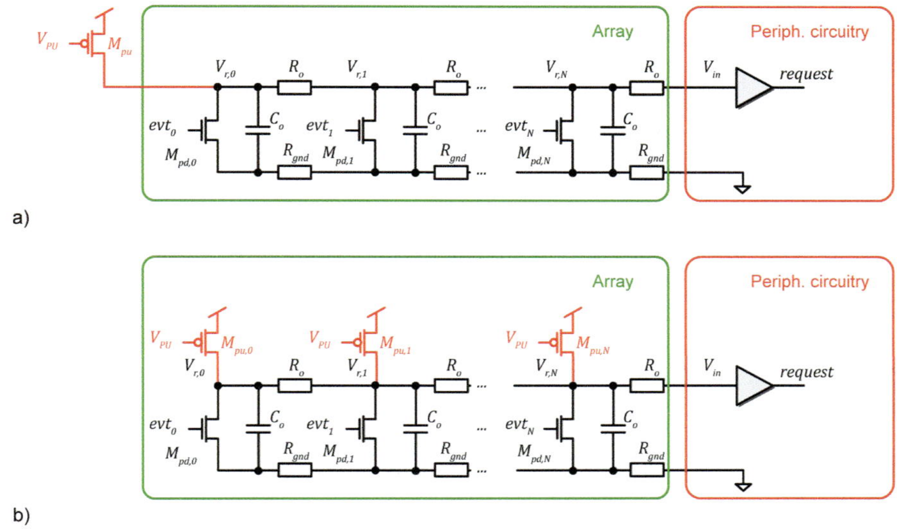

FIGURE 3.13 Techniques for mitigating voltage drop at the request line: a) Relocate the pull-up transistor. b) Split the pull-up transistor.

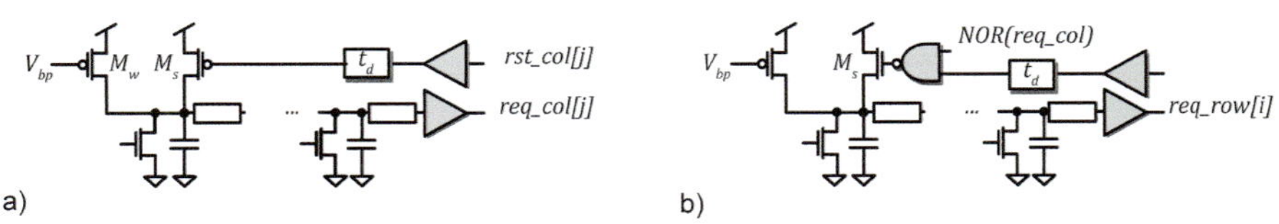

FIGURE 3.14 Implementation of adaptive pull-up elements a) at the column and b) row level.

M_s is a strong transistor that pulls the line up after all events on that line have been readout. The bias current of M_w, I_w can be selected to be as low as the total leakage current of the pull-down transistors of the pixel array, reducing the amount of wasted energy.

In principle, due to the random nature of events, it may seem difficult to predict when all events have been attended and the signal must be pulled up again. However, it can be predicted by considering the steps of the AER protocol in Figure 3.4. First, a request is made per row, and once it is acknowledged, the request is repeated at the column level, where only the elements of the acknowledged row can participate. This means that the $req\_c[[j]$ must always be pulled up after $rst\_col[j]$ is asserted. On the other hand, $req\_r[i]$ can be pulled up once $rst\_row[i]$ is activated and no request is made per column, i.e., when all signals within the req_col bus are deactivated. Note that the condition just described is the NOR operation of all elements on req_col, which can be implemented outside the pixel array with a negligible increase in area consumption. Figure 3.14 shows the schematic of both pull-up elements.

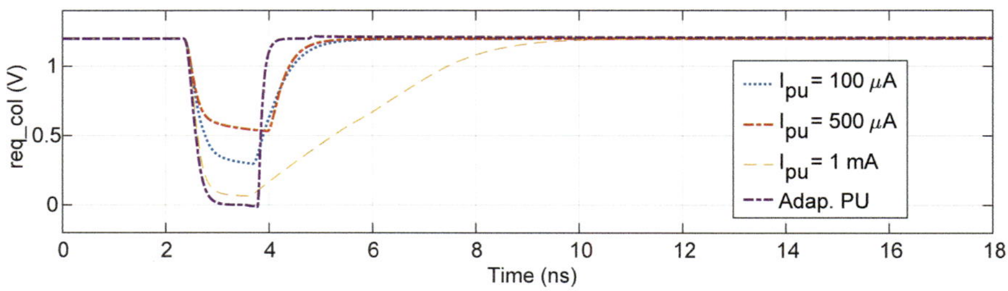

FIGURE 3.15 Simulation comparing adaptive and conventional pull-up elements.

A programmable delay is added after the reset signal to ensure the pixel removes the request before enabling M_s.

The operation $req\_col == N'b0$ refers to all bits of req_col logic high. Note that this modification not only improves the timing of the signals, but also reduces power consumption associated with pull-up elements.

Figure 3.15 shows simulation results using a PMOS as a pull-up element for different bias conditions and using the proposed adaptive pull-up element. The rise time of the signal increases with the strength of the transistor; however, the logic-low voltage also does. On the contrary, the adaptive pull-up element improves both metrics.

3.5 Programmable delay line

A programmable delay is crucial to implement time-gating techniques. As explained in Section 1.2.3, these techniques rely on a short-length gate that enables the pixel operation. This gate is shifted with respect to the laser pulse by a time Δt defining N_{bins} equally spaced bins. Finally, the ToF can be extracted from the rise/fall time of the histogram [Mori20a] or by means of the CoM algorithm [Gyon23; Rocc20].

The former method demands less computational effort and exhibits a well-defined resolution of Δt. However, for the circuit to work properly, laser pulses must be shorter than this resolution. Otherwise, the histogram will show the convolution of the enable gate and the laser pulse, causing smooth transitions in the histogram where the rise/falling time is difficult to extract.

On the other hand, the CoM algorithm has emerged as the preferred choice in recent studies [Gyon22]. It allows relaxing laser pulse specifications and enables sub-bin resolution, thereby allowing using longer values of Δt and a reduction in the number of exposed bins during a measurement. Thus, time resolution requirements must be evaluated prior to the implementation of the programmable delay line.

3.5.1 Time resolution requirements

The time resolution requirements are intricately tied to the depth resolution targeted by the sensor. While the performance of the laser pulse plays a crucial role, it is equally vital to ensure that the receiver is optimized to prevent becoming the limiting factor in measurements. Assuming an ideal laser pulse with a picosecond or even femtosecond pulse width, the resolution of the delay line can be derived from the speed of light, c.

However, in the case of implementing the CoM algorithm, the depth and time resolutions are no longer directly proportional due to this constant.

In this scenario, ToF is estimated as:

$$\frac{ToF}{\Delta t} = \frac{\sum_{i=0}^{N_{bins}} i\left(C_i - C_b\right)}{\sum_{i=0}^{N_{bins}} i\left(C_i - C_b\right)} \tag{3.5}$$

where C_i and C_b are the counts of each bin, and the counts due to background illumination, respectively. C_b is typically derived from the bin associated with the lowest counts. Despite initial appearances of increased computational complexity and potential slower processing due to the need to detect C_b prior to computation—requiring scanning all bins—this is not entirely true. By dividing terms in equation (3.5):

$$\frac{ToF}{\Delta t} = \frac{\sum_{i=0}^{N_{bins}} iC_i - 0.5N_{bins}\left(N_{bins} + 1\right)C_b}{\sum_{i=0}^{N_{bins}} C_i - N_{bins}C_b} \tag{3.6}$$

the ToF estimation can be performed in real time after each bin iteration, allowing for the background illumination level to be extracted at the end of the operation, as it is consistently multiplied by constants.

To evaluate the increase of resolution from Δt, we must assess the smallest detectable variation in depth by the sensor. Assuming that the depth variation is sufficiently small, and hence that power of the received light pulse is almost constant, we can infer that the total number of photons received by the detector remains constant. This ensures that the denominator in equation (3.5) remains constant and equal to the total number of counts C_T. Consequently, the minimum detectable variation in depth by the sensor corresponds to one photon transitioning from bin i to bin i + 1. As a result, the difference between these two measurements, ΔToF, can be expressed as:

$$\frac{\Delta ToF}{\Delta t} = \frac{i\left(C_i^{'} - C_i\right) + \left(i+1\right)\left(C_i^{'} - C_{i+1}\right)}{C_T - N_{bins}C_b} \sim \frac{1}{C_T - N_{bins}C_b} \tag{3.7}$$

where $C_i^{'}$ and $C_{i+1}^{'}$ represent the counts of the new measurement associated with bins i and i + 1, respectively. From equation (3.7) it can be deduced that the improvement in resolution is given by the SNR of the received signal. This suggests that both the distance and C_b degrades this factor, since SNR is reduced in both cases. Also, reflectivity of the target degrades resolution.

Thus, if a depth resolution Δd is to be achieved, the required Δt can be estimated as:

$$\Delta t \simeq \frac{\Delta d}{c} \cdot \frac{1}{C_T - N_{bins} C_b} \tag{3.8}$$

From equation (3.8) the energy of the laser pulse, E_L can be selected considering its wavelength and the background rejection level of the scene. However, this estimation is not trivial, since all photons impinging the active region are not detected and C_T strongly depends on the architecture of the pixel, the field of view of the emitter and its dynamics.

3.5.2 Circuit implementation

In practice, there exist multiple methods to implement a delay element in standard CMOS technologies. Figure 3.16 (a) shows the simplest implementation, where N_{inv} inverters are cascaded one after another. Each inverter presents a propagation delay, t_d, as shown in Figure 3.16 (b). This delay depends on technology parameters, geometry of devices, and parasitics. Sakurai's model [Saku90] estimated t_d as:

$$t_d \simeq \frac{\tau_{d,LH} + \tau_{d,HL}}{2} \tag{3.9}$$

$$t_{d,LH} = \left(\frac{1}{2} - \frac{1 - V_{t,p} / V_{DD}}{1 + \alpha} \right) t_t + \frac{C_L V_{DD}}{2 I_{DO,P}} \tag{3.10}$$

$$t_{d,HL} = \left(\frac{1}{2} - \frac{1 - V_{t,n} / V_{DD}}{1 + \alpha} \right) t_t + \frac{C_L V_{DD}}{2 I_{DO,N}} \tag{3.11}$$

Therefore, t_d depends on two terms. The first one is associated with the transition time of the input signal, t_t, and depends on the threshold voltage of devices, V_t, the power supply voltage, V_{DD}, and the velocity saturation index α, which accounts for discrepancies with the square law [Saku90]. The second term is related to the driving

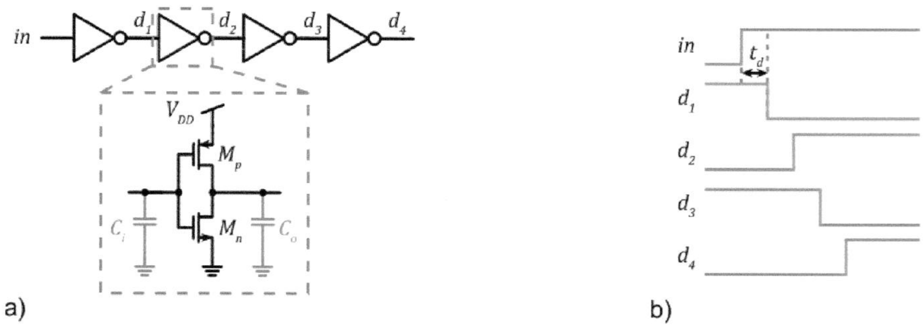

FIGURE 3.16 Delay line comprised of cascaded inverters: a) Circuit schematics. b) Waveform diagram.

capabilities of devices and depends on the load capacitance, C_L, V_{DD} and the current when the gate-source voltage and drain-source voltage of devices is maximum, I_{DO}.

Although Sakurai's model may not be as accurate to implement a high-resolution delay line, it can serve as a starting point, from which the circuit can be designed by using computer-assisted tools that implement advanced models of devices. Also, equations (3.10) and (3.11) provide intuition on how this delay can be tuned. For instance, t_d can be tuned by varying V_{DD} (also called, supply modulation), the driving current (referred to as current-starved inverters), or the load capacitance; however, its minimum value is limited by the intrinsic capacitances of the devices and, therefore, by the process technology. Another drawback of this implementation is that any asymmetry found in the physical implementation translates into mismatch in C_L of each stage, and thus, in asymmetries in t_d.

Additionally, using t_d as Δt would lead to reduced resolution and heightened sensitivity to both V_{DD} variations and jitter contributions across all stages. However, if the circuit does not require all phases simultaneously but employs them sequentially, Δt can be achieved by adjusting the value of C_L in each measurement. Figure 3.17 shows a circuit that enables a high-resolution delay where C_L is composed of N_c unit capacitors with capacitance C_o. Unlike the circuit from Figure 3.16 (a), this circuit implements a single delay phase upon the rising edge of V_{in}. Thus, the input-output delay is expressed as:

$$t_{del} = \frac{N_c C_o}{N_I I_b}\left(V_{ref} - V_{th}\right) = N_c \Delta t \quad (3.12)$$

where N is the ratio in the current mirror, V_{ref} is a reference voltage, and V_{th} is the threshold voltage of the comparator. From equation (3.12) it can be deducted that Δt can be adjusted varying any of these values. Since power-supply voltages are usually low in modern technologies, adjusting the value of C_L is the most versatile solution since a low resolution can be achieved. Additionally, adjusting N_I allows a coarse tuning of Δt.

The range of the programmable delay is thus limited by the number of unit capacitors of the circuit. However, it can be further extended by using clock phases coming from a low-jitter phased-locked loop [Suh20]. In fact, previous works in literature have demonstrated how pixels can individually select which phase of the clock to use as the input of a delay line depending on the depth ranges [Gyon23;

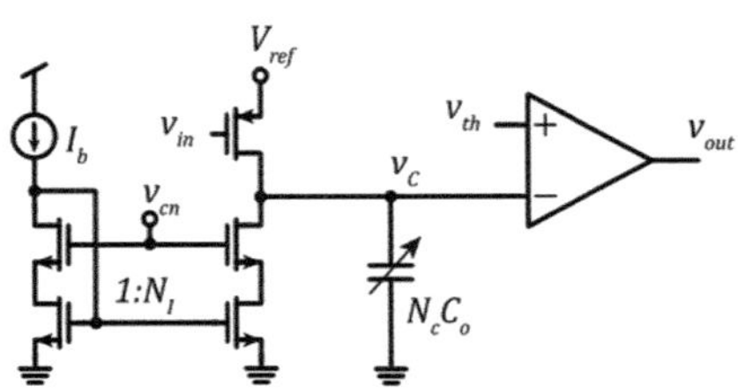

FIGURE 3.17 High-resolution delay based on a programmable capacitance.

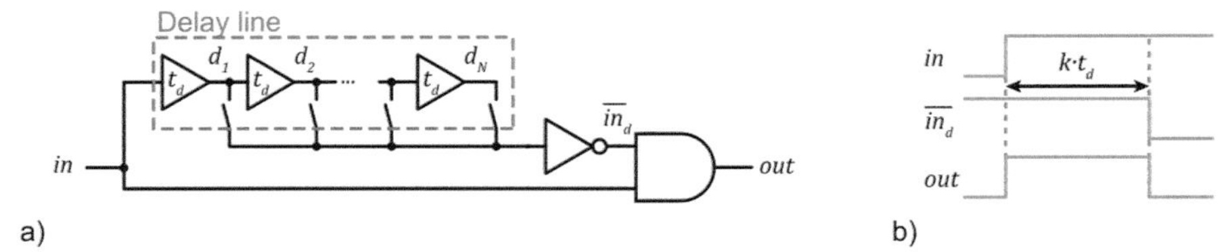

FIGURE 3.18 a) Shaping circuitry. b) Resulting waveforms.

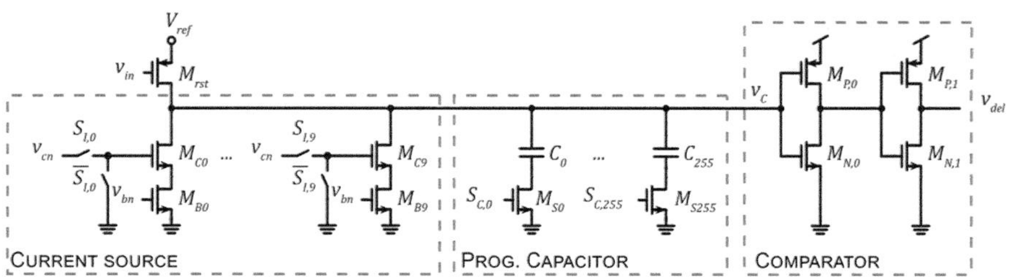

FIGURE 3.19 Programmable delay implemented in the SAER vision sensor.

Rocc20]. This aligns with the concept of histogram sliding described in Section 1.2.3, at the expense of an increase in pixel circuitry and interconnections.

Another critical aspect is the implementation of the comparator in Figure 3.17. Although adjusting the value of V_{th} may be useful to compensate for process variations, in practice, an inverter-based architecture may be of interest. This is because a fewer number of components results in fewer components contributing to noise and, therefore, less jitter in the measurement. Additionally, noise coming from V_{th} would be reduced. However, this comes at the expense of a fixed and process-dependent value of V_{th}. The subsequent section will explore this topic in greater depth.

Besides, the width of the output pulse must be shortened. This can be accomplished by using two phases from the circuit in Figure 3.16. Figure 3.25 depicts the implementation of the shaping circuitry. This circuit offers programmable functionality to select a specific delay phase, defining the pulse width of the output signal as equal to kt_d, where k represents the selected phase.

The SAER vision sensor implements a combination of the circuits from Figure 3.17 and Figure 3.25 to implement the global delay that is employed for ToF computation. Figure 3.19 depicts the schematics of the delay element, which precedes the shaping circuitry. The cascaded current source is composed of ten branches that can be individually enabled for coarse tuning. The programmable capacitor comprises 255 unitary capacitors with a common-centroid layout. This allows for 255 programming bins. While not all these bins are swept during a measurement, they facilitate compensating for any delay in the operation, such as the turn-on delay of the laser or compensate for temperature variations in the reference current. Also, note that the only source of mismatch that affects the operation is the mismatch in the unit capacitors. Consequently, this results in a versatile circuit capable of covering a wide range of delays.

Furthermore, the shaping circuitry comprises 32 buffers acting as delay elements. Each element induces a delay of approximately 430 ps, with a minimum delay of 2.4 ns. The choice of the minimum delay was determined by the constraints imposed by the impedance of the trace transmitting this signal to the pixel array.

3.5.3 Error sources

Even though equation (3.12) may suggest that Δt can be reduced to extremely low levels, this is not true in practice. Various error sources arise in the generation of the Δt that limits the minimum implementable value. Several effects significantly contribute to these limitations, in particular:

- Non-zero interconnection resistance and parasitic capacitance at the programmable capacitor.
- Impedance mismatch among unitary capacitors.
- Jitter at the output of the circuit.

Figure 3.20 depicts the model of the delay line including parasitics, where C_o is the unit capacitor, R_o the interconnection resistance and C_p is the parasitic capacitance modeled as a unit parasitic capacitance distributed across the entire circuit. In this model, capacitors are assumed to be connected in one dimension.

The total parasitic capacitance sets a minimum value for the delay, while the parasitic resistance limits dynamics by impeding the slope at the inverter from aligning promptly with the ideal case. This delay in the slope arises due to interactions among the RC elements, creating a deviation from the anticipated behavior.

Furthermore, since each element sees a different equivalent resistance, Δt becomes vulnerable to non-linearities due to a non-uniform Δt for different digital codes, potentially resulting in reduced resolution. To counter this issue, capacitors should be coupled to the programmable capacitor in pairs, where a capacitor at the start is linked simultaneously with a capacitor from the end of the line. By doing so, the impedance perceived by each capacitor element within the bank aligns, thereby augmenting the linearity of Δt. In the case of a two-dimensional array, this objective is accomplished by connecting four capacitors for each digital number, repeating this process in the orthogonal dimension.

Jitter, conversely, presents as a random deviation in the delay at the output. While histograms function as an averaging tool, diminishing the impact of jitter on the measurement, it can potentially compromise the performance of the system, even

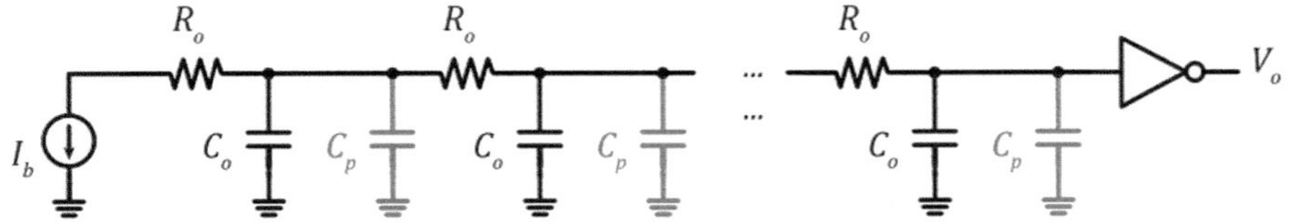

FIGURE 3.20 Delay line model including parasitics.

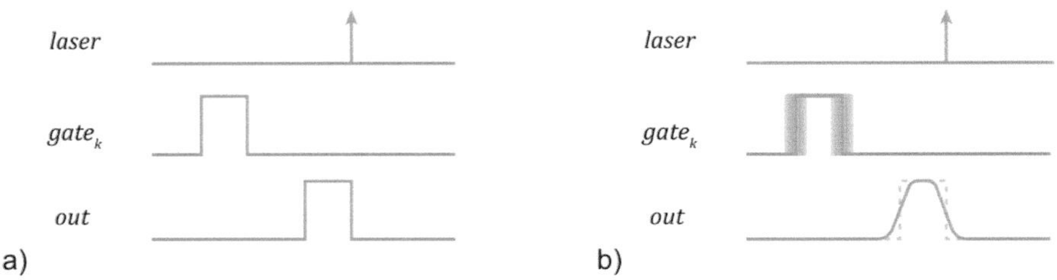

FIGURE 3.21 Reflected light pulse, gate of bin k, and output histogram from a) ideal gate, b) gate including jitter.

increasing in the number of exposures required for each bin. Figure 3.21 (a) illustrates the histogram reconstructed using a noiseless gate and a laser pulse with negligible width. On the other hand, Figure 3.21 (b) illustrates the effect of jitter, assuming a normal distribution. Since the gate position varies from sample to sample, jitter causes a smoothing effect on the rising edge. This smoothing implies that when the jitter becomes excessive, ToF extraction from the rising edge becomes impractical. While the CoM technique theoretically maintains its applicability without degradation, this assertion holds true only when a substantial number of samples are employed, i.e., after the collection of sufficient statistical data. Consequently, minimizing jitter is imperative to optimize measurement accuracy.

Noise analysis is commonly employed to model jitter and its dependence on circuit parameters. Within the circuit depicted in Figure 3.19, four discernible contributions of noise sources can be categorized:

- kTC noise stemming from the reset operation.
- Integrated noise attributed to the current source.
- Noise generated by the comparator.
- Power-supply-induced jitter.

A method to study the jitter contribution of an element involves estimating the integrated noise at the input of the comparator and then multiplying it by the relationship between this voltage and the delay, i.e., the slope of V_C, which can be extracted from equation (3.12).

Reset noise

The programmable capacitor is initially reset to V_{Ref}. Due to the utilization of low capacitance values, the impact of kTC noise becomes notable. This noise depends on the Boltzmann constant, k, temperature, T, and the value of the capacitance, C. However, despite this challenge, the signal in the voltage domain is represented by

$V_{ref} - Vth$, providing a high SNR with respect to this noise source. Nonetheless, it remains pivotal to carefully handle noise sources to attain optimal circuit performance. The jitter contribution of kTC noise in this specific circuit is expressed as:

$$\sigma_{t,rst}^2 = \frac{kTC_L}{\left(N_I L_b\right)^2} \tag{3.13}$$

NOISE CONTRIBUTION OF THE CURRENT SOURCE

Figure 3.22 shows the small-signal model when the comparator is about to toggle its output. This model is employed for the calculation of the noise arising from the current source and the comparator. For simplicity, only thermal noise is considered in this analysis. Although flicker noise is not negligible and causes discrepancies between analytical and electrical models, the tendencies developed in this analysis help understand how different parameters affect the jitter contribution.

Assuming that C_L is large enough, C_{gd} can be neglected, as most of the current will flow through the former. Also, the noise bandwidth is inversely proportional to the integration time, T_{int}, the integration time and can be approximated as t_{del}. Therefore, the integrated input noise associated with the current source, $\sigma_{t,i}^{\ 2}$, results:

$$\begin{aligned} \sigma_{t,I}^2 &\simeq 4kT\gamma N_I\left(g_{m,ib} + g_{m,ic}\right)\frac{C_L\left(V_{ref} - V_{th}\right)}{\left(N_I I_b\right)^3} \\ &= 4kT\gamma N_I\left(g_{m,ib} + g_{m,ic}\right)\cdot\frac{t_{del}}{\left(N_I I_b\right)^2} \\ &= \frac{8kT\gamma}{V_{dsat}}\frac{t_{del}}{N_I I_b} \end{aligned} \tag{3.14}$$

where $g_{m,ib}$ and $g_{m,ic}$ are the transconductances of the current source and cascode transistor, respectively, and γ is a technology constant. It is important to note that an ideal, noiseless current reference is assumed. However, in practical designs, the noise of the reference is added to this contribution. Equation (3.14) suggests that, for a given t_d, the noise contribution of the current source is only dependent on the current. Since transconductance increases as the square root of the current, noise decreases when increasing the current. It aligns with the conventional power-noise trade-off commonly encountered in electronics. Nevertheless, increasing the current requires a larger value of C_t to maintain t_d constant.

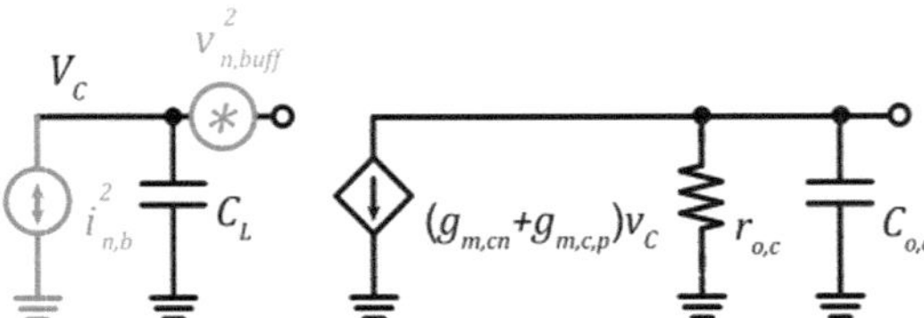

FIGURE 3.22 Small-signal model of the delay line in the SAER vision sensor.

Noise contribution of the comparator

The comparator is the last stage in the programmable delay line and has an impact on the accuracy and jitter of the output signal. When the buffer transitions its state, i.e., its input is near to V_{th}, both the PMOS and NMOS operate in the saturation region and its small-signal model can be used to study its noise contribution. The transconductance of the PMOS and NMOS is assumed to be the same and equal to $g_{m,}$ and $g_{m,cn}$, respectively. As a result, the equivalent integrated noise corresponding to the comparator, $\sigma_{t,c}{}^2$, can be expressed as:

$$\sigma_{t,c}^2 = \frac{\gamma}{A_{o,c}} \frac{8kT}{C_{o,c}} \left(\frac{C_L}{N_I I_b} \right)^2 = \frac{\gamma}{A_{o,c}} \frac{kT}{C_{o,c}} \left(\frac{t_d}{V_{ref} - V_{th}} \right)^2 \quad (3.15)$$

where $A_{o,c}$ and $C_{o,c}$ are the open loop gain and the output capacitance of the comparator, respectively.

From equation (3.15), it becomes evident that several strategies can be employed to minimize noise in the comparator. Initially, augmenting its voltage gain diminishes the input-referred noise. However, this implies reducing the bandwidth, potentially leading to distortion, as later elucidated. Additionally, increasing the output capacitance reduces integrated noise, albeit requiring increased current for a constant bandwidth. Hence, the optimal strategy, if viable, entails augmenting the voltage sweep to enhance the SNR. Nevertheless, this approach presents limitations when scaling the power supply voltage.

Power-supply-induced Jitter

Furthermore, when employing a CMOS inverter as the comparator, a notable concern is power-supply-induced jitter. Despite its area-efficient class-AB behavior, which outperforms in power consumption and bandwidth, noise in the power supply gets amplified by the PMOS device, thereby deteriorating noise performance. As a result, power-supply-induced jitter emerges as the predominant noise source in this implementation and can be modeled as:

$$\sigma_{t,vdd}^2 = \sigma_{vdd}^2 \left(\frac{g}{g_{m.p} + g_{m,n}} \right)^2 \left(\frac{t_d}{V_{ref} - V_{th}} \right)^2 \quad (3.16)$$

Comparator's bandwidth

While adopting an NMOS common source with an active load may enhance the noise performance of the circuit, it could introduce non-linearity across different bins. This arises due to the finite bandwidth of the comparator. When the frequency of the input signal surpasses the bandwidth, altering the slope of the input signal leads to varying effective gain across bins. This variability can be interpreted as a dependency of V_{th}

on the ramp slope, consequently causing non-linearities in Δt. To analyze this effect, we can model the comparator as a one-pole system with a bandwidth ω_o. Assuming the input is a ramp and utilizing the inverse Laplace transform, the output of the comparator $v_{o,}(t)$ manifests as follows:

$$v_{o,c}(t) = A_{o,c} \frac{N_I I_b}{C_L}\left(t - \frac{1 - e^{-t\omega_o}}{\omega_o}\right) \tag{3.17}$$

Current source's output resistance

Another source of distortion appears at the output resistance of the current source, $r_{o,}$. Although the previous analysis assumes a constant slope at V_c, and therefore, an infinite output resistance, such an ideal current source is impossible to implement. In practice, the output resistance causes non-linearities in the delay characteristics, especially when the voltage swing is large. To account for this effect, circuit equations can be solved assuming that the circuit is linearized around $V_c = V_{th}$. Thus, the voltage across the capacitor results:

$$v_c(t) = V_{ref} - N_I I_b \gamma_{o,i}\left(1 - e^{\frac{-t}{\gamma_{o,i} C_L}}\right) \tag{3.18}$$

Consequently, the time it takes for the capacitor to reach V_{th}, $t_d^{'}$, yields:

$$t_d = -\gamma_d C_L \quad \ln\left(1 - \frac{V_{ref} - V_{th}}{N_I I_b \gamma_{o.c}}\right) \tag{3.19}$$

Although equation (3.19) demonstrates a linear correlation between $t_d^{'}$ and C_L, it is important to note that the impact of PVT variations on $r_{o,}$ can result in inaccuracies in applications where precise delay values are crucial. Thus, to minimize this effect, a cascode structure is used in the design. To mitigate the effect of the output resistance, it is necessary to ensure that $t_d^{'}$ is significantly smaller than $r_{o,L}$, resulting in the following design equation:

$$r_{o,i} >> \frac{V_{ref} - V_{th}}{N_I I_b \left(1 - e^{-1}\right)} \tag{3.20}$$

Finally, if techniques such as dynamic element matching are implemented, systematic errors such as mismatches in the capacitor array can be randomized. In this scenario, the effect of mismatch is considerably reduced, but at the expense of an increased jitter. However, as the system is oversampled, this situation is preferred.

3.6 BIAS GENERATOR

Voltage and current references are pivotal components in high-performance ICs, significantly impacting system accuracy by establishing a base for measurements and biases for the circuits to operate in the desired operating point. The references should exhibit two essential characteristics: they need to minimize noise while also being resilient against fluctuations in the power supply. This requirement is not restricted solely to DC variations; it extends to any power supply noise, demanding a high *Power Supply Rejection* (PSR) to ensure that such noise is not coupled to the reference.

While voltage references must be insensitive to PVT variations, current references specifications can often be relaxed. In fact, since the transconductance of the MOS transistors (and therefore, the bandwidth of the circuits) decreases with the temperature, *Proportional-To-Absolute-Temperature* (PTAT) current references are usually preferred in circuit design [Raza17].

Furthermore, internally generating these references couples the signal to the reference of the IC (typically the substrate). Consequently, using external references may compromise the performance of the system. As depicted in Figure 3.23, employing an external voltage as a threshold for a comparator can exhibit an undesired behavior. While the external reference may remain stable and present a low-noise level with respect to the PCB ground, fluctuations in the IC ground might lead to a noisy internal reference. This scenario holds particular significance in image sensors that utilize linear-mode photodiodes, given their reliance on the IC substrate as a reference.

Figure 3.24 (a) provides an illustration of a PMOS current source biased using an external voltage. This circuit is not practical for two reasons. First of all, the bias

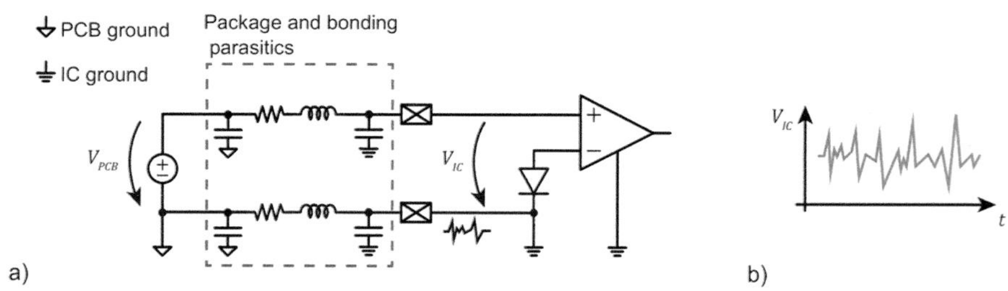

FIGURE 3.23 a) Noisy reference due to substrate noise. b) Voltage reference seen from IC ground.

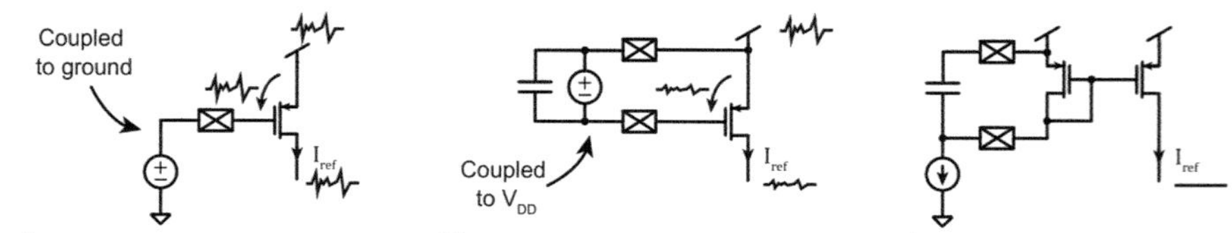

FIGURE 3.24 Biasing a current source with: a) non-coupled external voltage. b) External voltage coupled to V_{GS}. c) External current and decoupling capacitor.

voltage lacks an accurate relationship with the current making it challenging to establish an accurate correspondence between an absolute voltage and the actual current. Secondly, the external voltage is coupled to ground. Therefore, the current might exhibit a high noise level due to a poor PSR, as its gate voltage is not coupled to the source, thereby amplifying noise.

This can be improved by using a decoupling capacitor as in Figure 3.24 (b), but the circuit is still susceptible to variations on the internal power supply node and the current value is unknown. Consequently, external references employing a current bias as shown in Figure 3.24 (c) are preferable. In this setup, a known current exists in the circuit, and a current mirror ensures a constant gate-source voltage. However, for enhanced system accuracy, internal references are the preferred choice as they have the same reference as the IC and minimize the need for additional external components. This reduction in external components can positively impact both the size and costs of the system.

3.6.1 Current Reference

There exist many alternatives to generate a current reference in an IC. Figure 3.25 show three different current generators in standard CMOS processes. The circuit in Figure 3.25 (a) generates a reference current, I_{ref}, dependent on the value of the resistor R_{bias} and the gate-source voltage of the MOS transistor, V_{gs}:

$$I_{ref} = \frac{V_{DD} - V_{gs}}{R_{bias}} \tag{3.21}$$

Nevertheless, this implementation presents a high dependence on the supply voltage and a low PSR. Figure 3.25 (b) depicts the well-known beta-multiplier current reference [Liu98]. In this circuit, the PMOS current mirror forces the same current in both branches. If M_{n2} is sized K times wider than M_{n1}, the difference in the gate to source voltage of both devices drops across R_{bias}. Neglecting body effect, the reference current can be defined as [Liu98]:

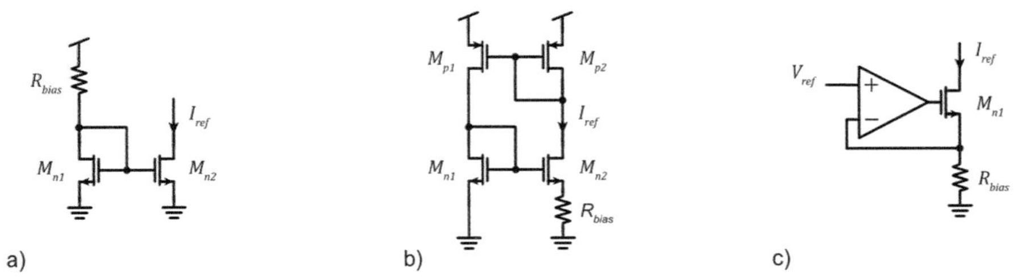

FIGURE 3.25 CMOS current references. a) Simple current reference using resistor and current mirror. b) Beta-multiplier current reference. c) Current conveyor.

$$I_{ref} = \frac{1}{R_{bias}^2 \beta_1}\left(1 - \frac{1}{\sqrt{K}}\right)^2 \quad (3.22)$$

where β_1 is the large-signal transconductance of M_{n1}. This current reference is also process dependent, but still qualifies for different applications where the reference current allows a certain margin. However, an interesting aspect of this circuit is that the transconductance of M_{n1} is fixed by R_{bias} and K, which is the parameter of interest when biasing a transistor. Both circuits generate a PTAT current, making these architectures suitable for biasing circuits such as amplifiers.

For applications requiring accurate current values unaffected by temperature variations, an optimal approach involves generating a current via a voltage reference, as depicted in Figure 3.25 (c). In this configuration, a feedback loop fixes V_{ref} across R_{bias}. Consequently, if the resistance is independent from temperature, the output current remains unaffected by temperature changes. However, it is worth noting that most resistors, particularly those integrated into an IC, demonstrate temperature dependency. Additionally, the offset of the amplifier also exhibits sensitivity to temperature changes, potentially compromising the performance of the circuit.

As previously mentioned, all circuits from Figure 3.25 rely on the value of R_{bias}. Therefore, ±20% variations related to the absolute value of an internal resistor are expected [Hast05]. This can be mitigated using trimming techniques [Hast05] or a high-precision external resistor. The former option can be implemented by selectively open circuit metal or poly fuses [Wan20], fixing the value of the resistor in a post-production.

Figure 3.26 (a) shows a 4-bit trimmable resistor using fuses and trimming pads, while Figure 3.26 (b) shows a digitally controlled resistor. Trimming pads allow adjusting the resistor in a characterization stage. However, the number of trimming pads, and therefore, the device area, increases linearly with the number of bits, with a trade-off emerging between trimming accuracy and cost. Furthermore, the large capacitance of these pads may worsen the performance of the circuit in certain applications. On the other hand, digital trimming uses an on-chip register to adjust the resistance value. Systems without a non-volatile memory need to store this value outside the IC and program it whenever the system is powered on. Alternatively, one-time-programmable fuses or electronically programmable memories can be implemented to avoid the use of trimming pads [Drag20]. Note that in both schemes,

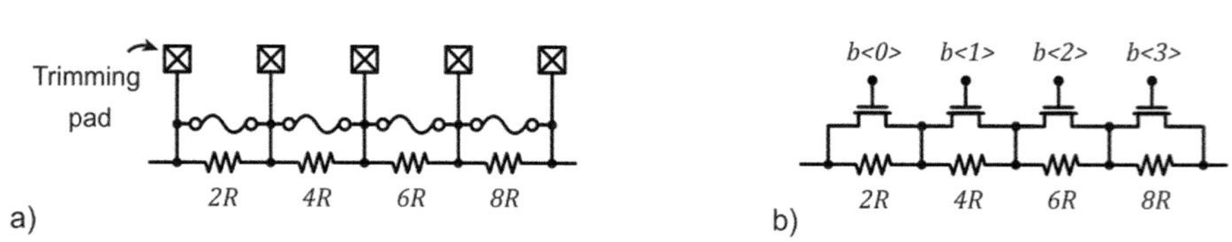

FIGURE 3.26 Binary-weighted resistor trimming using: a) Fuse trimming. b) Digital trimming.

a fixed resistor can be included in series to shift the absolute value of the resistance while achieving the same resolution.

The SAER vision sensor requires a stable current over temperature for the generation of the programmable delay presented in Section 3.5. Therefore, the circuit from Figure 3.25 (c) was implemented for that purpose. The circuit implements an integrated resistor with a fixed value of R_{fix} in series with a digitally controlled R_{trim}. The values of R_{fix} and R_{trim} as well as the number of control bits must be selected based on the desired range and accuracy of the final value of the resistor. For instance, if the circuit must account for a range of $\pm\alpha\%$, R_{fix} and R_{trim} must be selected such as:

$$\left(1+\frac{\alpha}{100}\right)R_{fix} = R_{target} \tag{3.23}$$

$$\left(1-\frac{\alpha}{100}\right)\left(R_{fix}+R_{trim}\right) = R_{target} \tag{3.24}$$

where R_{target} is the target resistance value. Furthermore, the number of bits for R_{trim}, N_R, is chosen considering the LSB, at the worst-case scenario, i.e., when the sheet resistance is maximum. To achieve an error lower than $\epsilon_R\%$, N_R, must be selected considering the following design equation:

$$\frac{1}{R_{target}}\frac{\Delta R_{max}}{2^N-1} < \frac{\varepsilon_R}{100} \tag{3.25}$$

3.6.2 VOLTAGE REFERENCE

Regarding voltage reference generation, bandgap references are the preferred option, because this circuit generates a well-defined temperature-independent voltage [Raza17]. To implement a temperature-independent voltage reference, the bandgap reference adds two voltages with opposite dependence on temperature, i.e., a *Complementary-To-Absolute-Temperature* (CTAT) voltage and a PTAT voltage.

Almost every component in an IC exhibits temperature dependence. Most of them are undesired, related to the materials employed for their implementation. Although temperature sensors based on variations in the resistance value of a poly resistor can be implemented [Choi18; Pan18], these are not accurate enough for voltage reference. On the contrary, CTAT and PTAT behavior can be extracted from device physics, which are much more reliant. Particularly, the forward bias of diodes exhibits a CTAT behavior [Raza17]. However, if the forward voltage of two diodes with different areas drawing the same current is subtracted, the thermal voltage, U_t, can be extracted [Raza17].

Figure 3.27 (a) depicts the basic concept of a bandgap reference. The forward biased diode decreases with the temperature while the U_t increases. Therefore, the temperature coefficient of the output voltage V_{ref}, can be compensated by adjusting

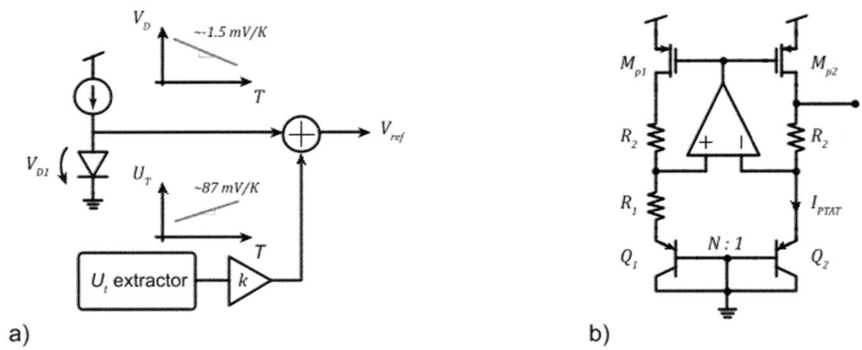

FIGURE 3.27 Bandgap reference. a) Concept. b) Typical implementation.

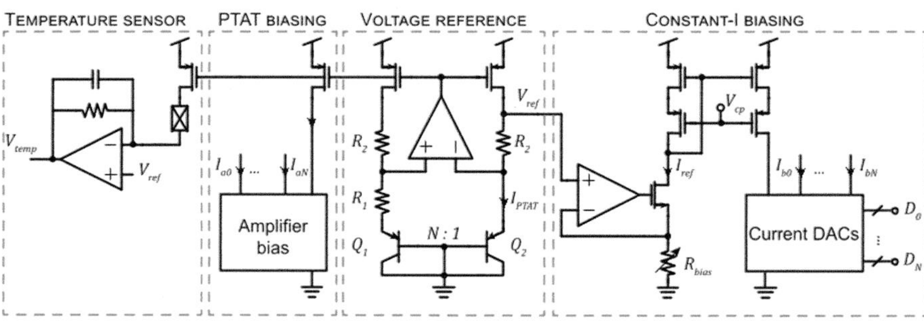

FIGURE 3.28 Simplified block diagram of voltage and current references in the SAER vision sensor.

the value of k. Figure 3.27 (b) shows one of the most common implementations of a bandgap reference. In this circuit, two parasitic *Bipolar Junction Transistors (*BJTs) with a ratio of N:1 are used as diodes. The feedback loop fixes the voltage across R_1 to the difference between the base-emitter voltage, V_{BE}, of both BJTs:

$$V_{R1} = V_{BE2} - V_{BE1} = U_t \ln(N) \tag{3.26}$$

Finally, the gain k from Figure 3.27 (a) is implemented using the voltage-to-current conversion in R_1 and the current-to-voltage conversion in R_3.

The voltage reference from the bandgap serves as the reference voltage for the circuit in Figure 3.25 (c). Additionally, the PTAT current generated by the bandgap is used to bias amplifiers in the sensor, ensuring a constant bandwidth for the circuits. As illustrated in Figure 3.28, the references utilized by the SAER vision sensor are generated in this manner. Programmable current mirrors are used to configure circuit parameters within the sensor, including setting the dead time of SPADs or adjusting the strength of pull-up devices in the AER readout block, among others. Also, a replica of the PTAT current is connected to an output pad to measure the temperature of the chip.

3.7 DIGITAL CORE

While the external enable signal regulates the operation of the sensor and the AER interface handles output data transmission, a dedicated digital circuit is responsible for controlling internal registers for configuration purposes. This circuit also manages tasks like programming the SRAM of the pixel array and handling test signals from the sensor.

Figure 3.29 displays a block diagram of the digital core of the SAER vision sensor, encompassing several key blocks:

- **SPI Interface**: Manages communication. The sensor acts as the slave.
- ***Finite-State Machine* (FSM)**: Handles and parses SPI commands.
- **Register bank:** Stores configurations of the sensor.
- **SRAM controller:** Generates the signals to update the SRAM memory of pixels. It is an extension of the SRAM controller presented in Section 3.3.2.

The proper operation of all these blocks is crucial to guarantee the performance of the SAER vision sensor. Since no on-chip processing is carried out in the digital part, this circuitry operates at a moderate frequency level. The maximum operating frequency was selected to be 20 MHz, to save area and power consumption. Also, the digital core multiplexes test signals in the digital test bus, $test\_d[3:0]$.

3.7.1 FRAME STRUCTURE IN SPI COMMUNICATION

Similar to other communication interfaces, the SPI manages the transmission and reception of bits in a synchronized manner. These bits are organized into bytes, forming the fundamental building blocks of data frames. However, issues such as data corruption, loss, or mismanagement due to incorrect command sequences could lead to errors or undesired behaviors. Hence, establishing a defined structure for

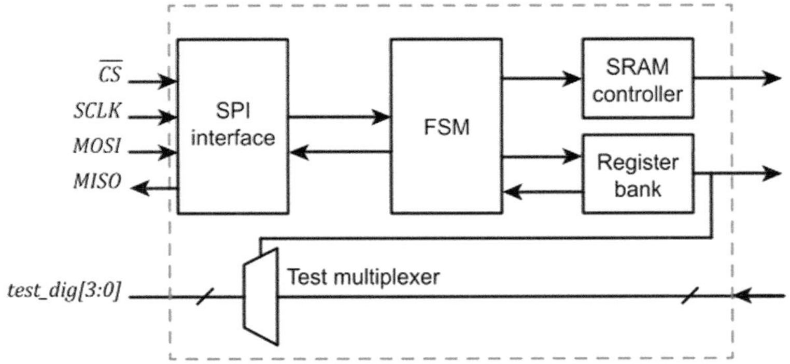

FIGURE 3.29 Simplified block diagram of the digital core of the SAER vision sensor.

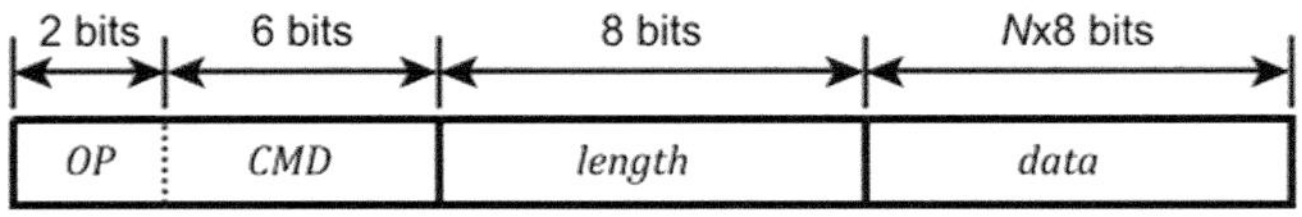

FIGURE 3.30 Structure of data frame in the SAER vision sensor.

a data frame is crucial for implementing a reliable and efficient protocol over SPI communication.

Figure 3.30 depicts the structure of the data frame in the SAER vision sensor. Each data frame must be composed of at least three fields. These fields correspond to:

- **Control byte.** This byte contains information with regard to the operation mode, OP, and the command, CMD. OP is ignored if the command is not related to a register.
- **Length byte.** The number of *bits* to be transmitted is sent as $length$. This byte is omitted if the operation is not related to a register.
- **Data byte(s).** Data associated with the SPI command, e.g., data to be written in a register. In case more than one byte is sent, data must follow a big-endian sequence.

The reason for sending the number of bits to be transferred will be explained in subsequent sections.

3.7.2 SPI commands in the SAER vision sensor

The SAER vision sensor incorporates an instruction dataset comprising numerous SPI commands used for updating internal registers, resetting specific blocks, or triggering the SRAM controller. At this moment, the data frame allows up to 64 different commands, yet only 24 of these commands are implemented in the current prototype.

Table 3.4 summarizes the SPI commands and their descriptions. Apart from the commands designated for writing and reading registers in the register bank, specific commands are involved in the operation of the sensor. Note that each command requires a specific number of bytes to be sent in the data field, and this will be elaborated on in subsequent sections.

Table 3.5 depicts available operation modes. The command identification as a write or read command is contingent upon the bits transmitted in the *OP* field. Additionally, simultaneous write and read operations can also be executed.

3.7.3 Configuration registers

The configuration registers are arranged as displayed in Figure 3.31. As the SPI frequency is more than 8 times lower than the system frequency, the sequential reading

TABLE 3.4
Available SPI commands in the SAER vision sensor.

SPI command	CMD (Hex.)	Data size (bytes)	Description
REG_XX	0x01-0x12	1-16	Reads/Writes data in register XX.
RST	0x13	2	Reset operation. Followed by reset command.
SRAM	0x14	1	Row/column reset signal to the pixel array.
WRITE_DELAY	0x15	1	Global input data for pixel SRAM.
HELLO	0x2A	0	Returns a predefined word. This word varies depending on *OP*.
FLUSH	0x3F	0	Flushes the FSM in case

TABLE 3.5
Operation modes in SPI commands.

OP	Mode	Description
00	Write & Return	Writes to register the information provided in the data field and returns the received data immediately after.
01	Write	Writes to register the information provided in the data field and does not send relevant information.
10	Read	Reads register data. Data reception starts from the third byte.
11	Read & Write	Reads register while writing incoming data.

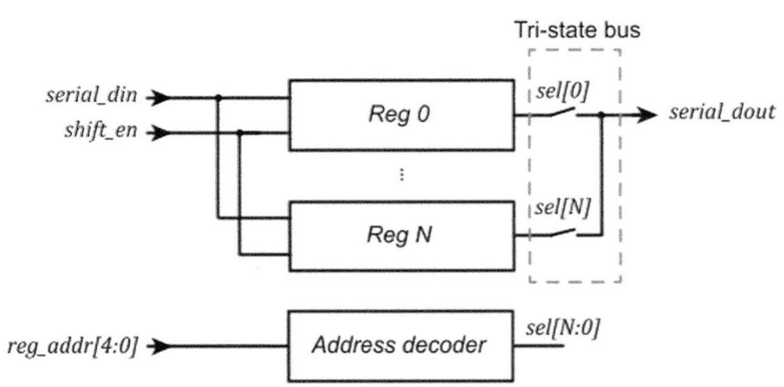

FIGURE 3.31 Interconnections in the register bank.

and writing of registers are employed to minimize bus congestion and reduce area consumption. This strategy allows the definition of registers with flexible sizes within a scalable architecture. The input and output descriptions of the register bank are provided in Table 3.6.

All registers share a common input, *serial_din*, for serial data input. During each clock cycle, the bits from register i shift when the *shift_en* signal is asserted and the corresponding signal *reg_[i]* is enabled. The *reg_[i]* signal is generated by the address decoder, converting the binary register address *reg_addr* into a one-hot encoding. When *reg_[i]* is activated, the MSB of the register is linked to *serial_dout*, operating as a tristate bus.

To ensure accurate write operation and prevent undesired states, each register cell includes an output register alongside the shift register. The output register is updated when the write operation concludes, indicated by the *reg_[i]* signal going low.

During read operation, *serial_dout* is connected to *serial_din*. To maintain bit alignment, the *length* field must specify the number of bits in the desired register. Contrarily, during register write operation, complete bytes are received. Consequently, the length field may contain more bits than the register size due to the redundancy of MSBs within the most-significant byte. In Figure 3.32, an illustration showcases

TABLE 3.6
Signals involved in the operation of the register bank.

Signal	Direction	Description
serial_din	Input	Serial data input from FSM.
serial_dout	Output	Serial data output from register bank.
shift_en	Input	Shift enable signal from FSM.
***reg_addr*[4: 0]**	Input	Address bus

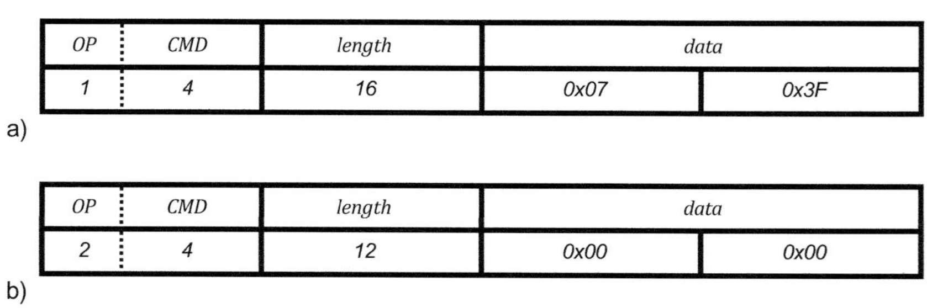

FIGURE 3.32 Sample data frame for interacting with a 12-bit register: a) writing and b) reading operation.

TABLE 3.7
Configuration registers available in the SAER vision sensor.

Register name	Addr.	Bits	Signals	Description
REG_AER_CFG	0x01	4	*bus_req_nand*	1: Nand tree for *bus_req_x/y* 0: Pull-up devices
		3	*latch_req_en*	1: Latches *req_row/col*
		2	*latch_rst_en*	1: Latches *rst_row/co*
		1	*din_sram_ext*	1: SRAM input data from CPM. 0: SRAM input data from digital core.
		0	*pu_boost_en*	1: Enables adaptive PU.
REG_DELAY_BIAS	0x02	9:0	*cfg_delay_bias*	Selects N_l in the programmable delay (thermometric)
REG_DELAY_CFG	0x03	3	*delay_bypass*	1: Bypasses the programmable delay
		2	*win_bypass*	1: Bypasses windowing circuitry
		1	*Unused*	-
		0	*delay_en*	0: Disable current sources in the programmable delay
REG_BIAS_MUX	0x04	4	*sel_pix_vbq*	1: Use external references.
		3	*sel_pix_vrst*	0: Use internal references
		2	*sel_aer_pd*	
		1	*sel_aer_pu_d*	
		0	*sel_aer_pu*	
REG_WIN_SIZE	0x0F	31:0	*cfg_win_size*	Configuration bits for the width of *enable* (one hot).
REG_SRAM	0x10	127:96	*sram_dig*[3]	Input data for the SRAM
		95:63	*sram_din*[2]	controller in the AER
		63:32	*sram_din*[1]	readout block.
		31:0	*sram_din*[0]	

writing and reading of a 12-bit register at address 0x04. For the write operation, two bytes are transmitted to update the value of the register to 0x73F, and the length *field* is set to 16. However, when reading the register, the length field needs to indicate the register size.

On the receiver side, bytes are received in a big-endian format regardless of the operation type.

Table 3.7 and Table 3.8 depict the available configuration registers in the SAER vision sensor, along with the different signals stored in each register. Registers not listed are reserved for debugging and testing purposes.

TABLE 3.8
Bias registers available in the SAER vision sensor.

Register name	Addr.	Bits	Signals	Description
REG_BIAS_RES	0x05	5	*sel_res*	1: Use external resistor 0: Use internal resistor
		4:0	*cfg_res*	Configuration bits of internal resistor
REG_BIAS_VPD	0x06	4:0	*cfg_aer_pd*	Configuration bits of AER pull-down transistors.
REG_BIAS_VPU	0x07	4:0	*cfg_aer_pu*	Configuration bits of AER pull-up transistors.
REG_BIAS_VPU_D	0x08	4:0	*cfg_aer_pu_d*	Configuration bits of delay in adaptive pull-up transistors.
REG_BIAS_VBQ	0x09	4:0	*cfg_pix_vbq*	Configuration bits of quenching bias.
REG_BIAS_VRST	0x0A	4:0	*cfg_pix_vrst*	Configuration bits of SPAD dead time.

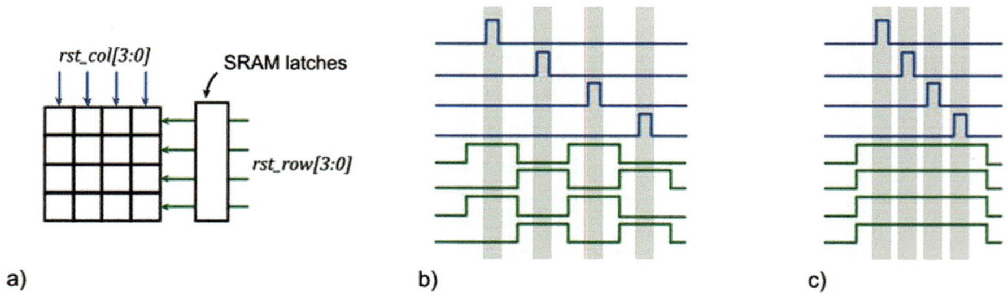

FIGURE 3.33 a) Signals involved in the write operation of the pixel SRAM. b) Waveform diagram of a column-wise operation. c) Waveform diagram of global write operation.

3.7.4 SRAM WRITE OPERATION

The digital core manages the writing process for the SRAM within the pixel array, generating essential signals for updating the SRAM of pixels via the AER readout block. Figure 3.33 (a) provides an overview of these signals. The signal *wr_sram*, which remains active throughout the operation, has been omitted for clarity. The operation of the SRAM controller depends on the byte received in the *data* field. Table 3.9 summarizes the different commands that the SRAM controller can manage.

Unlike the write or read operation of a register within the register bank, the SRAM controller requires more than one instruction to operate. The reason for that is the digital circuitry allows the user to control the timing of the operation. Therefore, when the circuit is in idle state, the first SRAM command corresponds to the writing mode and the rest correspond to operation commands, until all columns have been written or the user sends a termination command.

TABLE 3.9
SPI commands of the SRAM controller.

Command type	Byte	Description
Mode	0x00	Write SRAM of pixels column-wise.
Mode	0x01	Write SRAM of all pixels with the same value.
Operation	0x02	End operation and return to idle state.
Operation	0x0A	Enable *wr_sram_row* and *rst_row* depending on mode: -0x00: Enable all rows. -0x01: Toggle between activating even or odd rows.
Operation	0x0B	Deactivate *rst_row* and *wr_sram_row* and activate *rst_col* and *wr_sram_col*

Figure 3.33 (b) illustrates the signal flow involved in a column-wise write operation of the SRAM, using a 4×4 pixel array as an example. Initially, the SRAM command is followed by the *mode* byte. Subsequently, the SRAM command is re-sent alongside an operation byte, activating signals at the row level, highlighted in light gray. At this moment, SRAM latches are updated with the value stored in *REG_SRAM*. Following this, another operation byte triggers signals at the column level to write data into that specific column, shown in dark gray.

Upon successfully writing data to the odd pixels within the first column, the value stored in the *REG_SRAM* is updated to initiate the process anew. It is important to note that this sequence concludes once the last column is written. However, due to *rst_row* toggling from column to column, the entire procedure must be repeated, this time starting from the even rows by sending the command 0x0A twice at the beginning.

The operation simplifies when writing the entire array. Figure 3.33 (c) depicts this operation. First, information is updated at the row level and then a global reset operation is triggered to start a staggered writing operation, as mentioned in Section 3.3.2.

3.7.5 Controlling the Programmable Delay

The programmable delay is also managed via the SPI interface. This delay can be adjusted by either increasing or decreasing the number of capacitors by an arbitrary value ranging from 1 to 64. During this process, the data field comprises a single byte, which is divided into two sections: the operation (2 MSBs) and the quantity of capacitors to be added or subtracted (6 LSBs). Table 3.10 depicts the different operation codes and their functionality.

3.7.6 Reset Modes

In addition to configuration settings, the digital core provides the capability to selectively reset essential circuits as required. These circuits include:

TABLE 3.10
Operations in the control of the programmable delay.

Operation code	Description
2'b00	Increase the capacitance.
2'b01	Decreases the capacitance
2'b10	Sets the maximum capacitance.
2'b11	Sets the minimum capacitance.

TABLE 3.11
Reset modes.

Operation code	Description
0x01	Reset bandgap reference.
0x02	Reset AER circuit at the column level.
0x03	Reset AER circuit at the column level (staggered).
0x04	Reset AER circuit at the row level.
0x05	Reset AER circuit (row and column) and pixel array.
0x06	Reset SRAM latches and set pixel array to default.
0x07	Reset SRAM latches.

- Bandgap reference.
- AER readout block.
- Pixel array.
- Registers of the SRAM controller.

The reset command is followed by two bytes in the *data* field. The first one corresponds to the reset mode, explained in Table 3.11 and the second one to the number of clock cycles of the reset operation.

3.8 OPERATION MODES AND DATA FLOW

The SAER vision sensor offers different operations modes, each of those offering different advantages depending on the scenario and applications. Although it is intended to operate as an octopus retina, where all pixels generate information in a continuous manner [Culu03], alternative modes can be implemented depending on the timing of control signals and how data is managed by the external receiver.

3.8.1 Free-running mode

The default operation mode is the FR mode. In this mode, pixels are continuously enabled, after an initial reset. Pixels encode the intensity level of the scene into spike trains, as illustrated in Figure 3.34. These signals can then be used by the external receiver to reconstruct the signal of each pixel in real time.

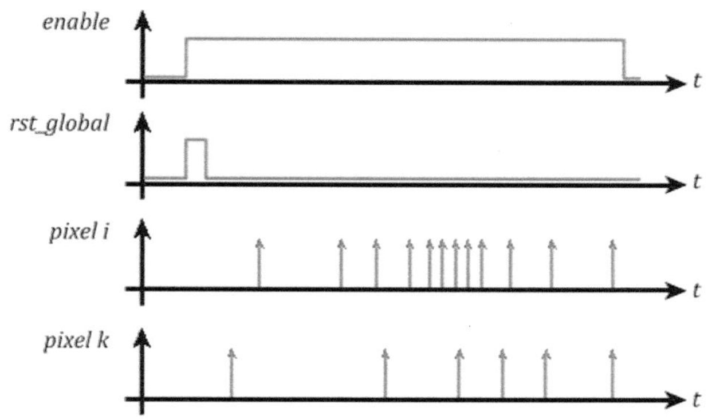

FIGURE 3.34 Waveform diagram of the sensor operating in Free-Running mode.

In FR mode, each event or spike is related to the detection of N_{ph} photons, and as explained in Section 2.5.2, the value of N_{ph} can be individually adjusted for each pixel. Thus, the acquisition rate of each pixel is closely related to the intensity level.

While intensity frames can be reconstructed by computing the average event rate per pixel within a defined time span, the primary advantage of this operational mode is the absence of pixel signal constraints imposed by a frame rate. Specifically, information is transmitted to the receiver once pixels gather sufficient statistical data, and their signals are no longer limited by photon shot noise. This capability, combined with a neuromorphic processor directly handling spiking data, facilitates rapid processing of visual information.

3.8.2 Time-to-N-Photons mode

Alternatively, the sensor can be configured so that each pixel transmits only a single event. This mode is akin to the conventional time-to-first-spike mode documented in literature and it is a particular case of the FR mode where pixels only generate one event.

Figure 3.35 illustrates the control signals and pixel outputs in TNP mode. Although not shown, *wr_sram* is enabled to update SRAM upon event readout, thereby deactivating the pixel in such a way that each pixel only spikes once. Consequently, the number of events sent to the external receiver is reduced.

The TNP mode, although somewhat similar to the traditional operation of conventional image sensors with limited exposure time and frame rate, presents several distinct advantages. Unlike the separation of exposure and readout into different phases in conventional sensors, TNP mode asynchronously generates and transmits data to the external receiver. This feature not only enables the receiver to start processing the visual scene immediately but also allows for the termination of exposure once sufficient information has been gathered, given that not all pixels are required to extract information from an image [Torr09].

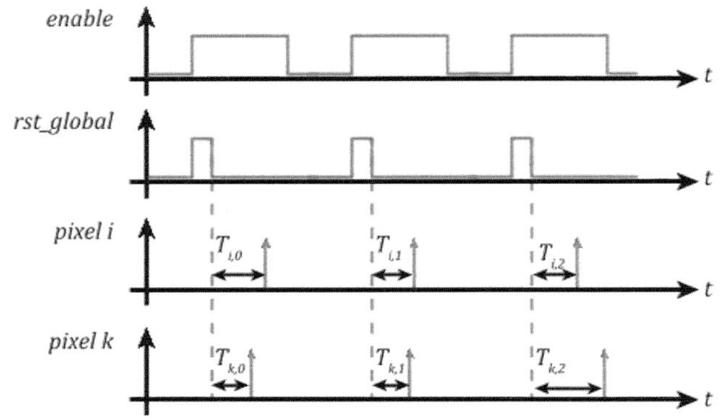

FIGURE 3.35 Waveform diagram of the sensor operating in Time-to-N-Photons mode.

Compared to the FR mode, this operation mode offers additional benefits. It produces less data, albeit at the cost of restricting the SNR of all pixel signals to N_{ph}, as information is reconstructed by a single spike. Moreover, the deactivation of pixels after readout prevents potential saturation of the periphery caused by a few highly illuminated pixels.

3.8.3 Quanta Imaging

The SAER vision sensor also enables a multi-bit QI mode [Foss13]. In this mode, the sensor is continuously gated, generating binary frames where logic 1 is related to the collection of N_{ph} or more photons. Then, these binary frames can be integrated to encode the intensity level, I, into the probability of detecting more than N_{ph} photons ($k > N_{ph}$) [Foss13]:

$$I = P\left[k > N_{ph}\right] = 1 - e^{H} \tag{3.27}$$

where H is the quanta exposure, i.e., the number of expected photons in the defined exposure time [Foss13].

Figure 3.36 shows how the sensor is operated in the QI mode. As in the TNP mode, pixels are disabled after readout to ensure that each pixel is only readout once during each exposure. Thus, the asynchronous readout performed during the exposure allows the sensor to reduce the dead time between exposures as much as possible. This fact is of particular interest since the effective data rate increases as almost no gap appears between binary frames if the exposure time is sufficiently long.

Finally, binary frames are sent to the CPM, which can either integrate the signals to reconstruct a frame or send the raw data to the next element in the processing chain.

3.8.4 Time-of-Flight mode

The ToF mode works in a similar way to the QI mode. However, the *enable* signal goes through the programmable delay. Thus, the signal sent to the pixel array is shortened

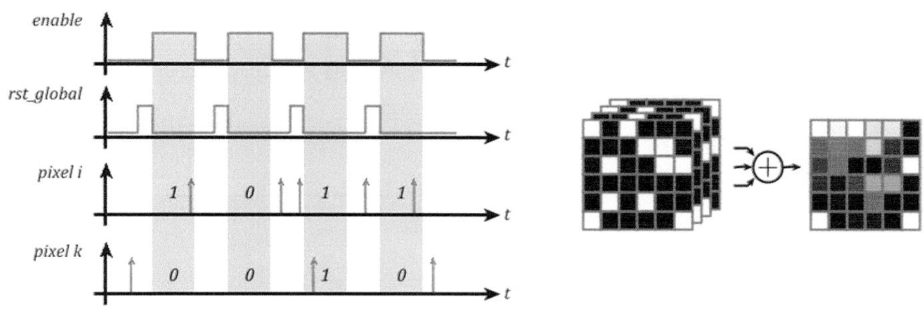

FIGURE 3.36 Waveform diagram of the sensor operating in Quanta-Imaging mode.

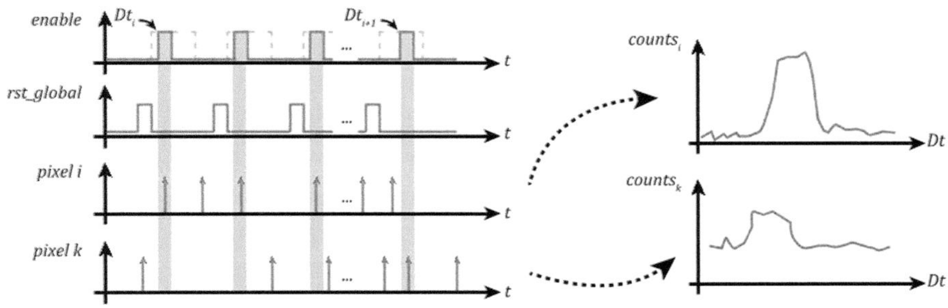

FIGURE 3.37 Waveform diagram of the sensor operating in Time-of-Flight mode.

and slightly delayed from the laser pulse, as shown in Figure 3.37. Pixels are exposed multiple times for each bin to generate histograms. These histograms can be limited to six gate positions in order to achieve a considerable acquisition rate [Rocc20].

In this mode, the exposure time of pixels is as short as 3 ns. Consequently, the readout is carried out after the exposure in a separate phase. However, only pixels detecting a photon transmit an event. Additionally, pixels detecting a photon do not require being read out after each exposure. By adjusting N_{ph}, it is possible to focus solely on gathering information from bins that receive a certain number of events, significantly reducing data transmission and the time required for readout. Following the initial event, N_{ph} can be set to 1 to gather information with photon resolution. This strategy can also prevent receiving information from pixels that have not accumulated sufficient statistical data to accurately compute ToF.

4 The SAER pixel

Image and vision sensors are modular systems consisting of pixel arrangements. Depending on the targeted sensor behavior, pixels are meant just to acquire light and perform some local error correction [Font11] or include some pre-processing capabilities seeking to handle information at the edge [Koch94; Rodr18]. The target in the former case is minimizing the number of transistors per pixel, even resorting to transistor-sharing strategies, and increasing the fill factor. However, vision sensors relax this specification and provide room for additional circuitry, empowering them to execute sophisticated functions and early-stage vision tasks.

Pixels within LiDAR-based vision sensors, on the other hand, carry intrinsic supplementary circuitry for tasks such as Time-of-Flight resolution through precision in-pixel TDCs [Brag14; Vorn17], photon counting [Dutt16], or in-pixel preprocessing, exemplified by in-pixel pre-histogramming techniques [Gyon23; Rocc20]. While quanta imaging [Foss13] and time-gated pixels [Mori20a; Panc13] offer promising avenues for acquiring both 2D and 3D information with minimal device count, they introduce an increment in the number of samples and subsequent acquisition time. As introduced in Chapter 3, the pixel featured in the SAER vision sensor embodies a multi-modal functionality for both 2D and 3D imaging, with its operational mode controlled at the algorithmic level.

4.1 PIXEL ARCHITECTURE

The architecture of the pixel serves as the bedrock upon which the functionality of the sensor rests, enabling a carefully designed interplay of components. This section explains the conceptual block diagram that underlies every pixel, describing the different elements that enable its multiple operation modes. From single photon receptors to event-driven readout logic, each component plays a pivotal role in capturing and processing light-induced signals. The implementation of the pixel, discussed in detail herein, unveils the inner workings of its integrated components, including the single-photon receptor along with its quenching and recharge circuitry, the photon integrator, the comparator for event assertion, and the AER readout logic.

 DOI: 10.1201/9781003427490-4

4.1.1 Conceptual Block Diagram

Figure 4.1 depicts the essential components of the SAER pixel, each represented as a discrete black box and contributing to the overall functionality of the pixel. These elements include:

- **Single Photon Receptor**: The single photon receptor stands as the initial point of contact for incoming photons. Upon photon arrival, this receptor swiftly translates the energy into a measurable electrical signal. This pivotal step forms the foundation for subsequent processing.
- **Photon Integrator**: The photon integrator assumes the crucial task of aggregating photon counts. Through its integration function, it accumulates discrete photon events, paving the way for a cumulative representation of the incoming light. Also, the photon integrator can serve as a memory to store the number of collected photons between exposures. Note that the operation of the integrator is orchestrated by the *enable* signal.
- **Comparator**: It is in charge of determining the generation of an event. The comparator asserts its output when the integrated signal surpasses a predefined threshold.
- **AER Readout Logic**: Interconnected with the readout circuitry described in Section 3.3.1, the AER readout logic implements the AER protocol, orchestrating the transmission of events outside the pixel array. It enables the flow of information, ensuring that the events are appropriately conveyed for subsequent processing stages. Also, it generates the internal signal *rst_pix* that resets the photon count.

While the SPAD front-end is inherently analog, the photon integrator and comparator can be implemented in either the analog or digital domain. Subsequent sections will describe the circuit implementation and possible alternatives, along with their advantages and disadvantages. This foundational structure establishes a robust basis for a wide array of applications, encompassing both 2D and 3D imaging techniques.

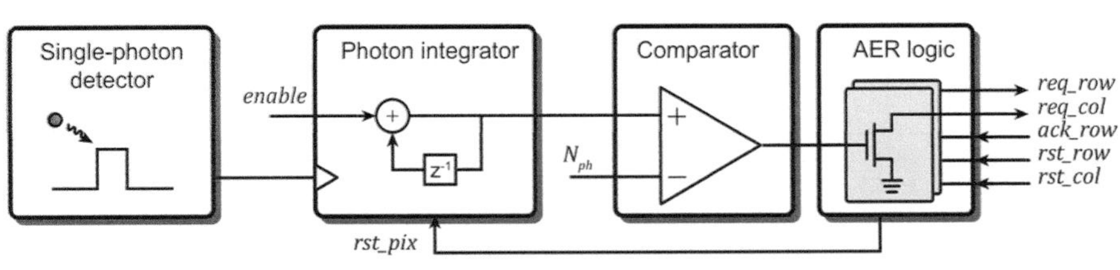

FIGURE 4.1 Conceptual block diagram of the SAER pixel.

4.1.2 Pixel implementation

Figure 4.2 complements the block diagram in Figure 4.1 by providing larger details of the actual circuit implementation, providing insights into the specific components that comprise the pixel architecture. The key elements within this implementation include:

- **SPAD detector**: Within the SPAD front-end, the active quenching and recharge mechanism stands as a critical element. It is responsible for extinguishing the avalanche process in the SPAD after a photon event, allowing for rapid recovery and subsequent signal processing. A configurable dead time parameter governs the duration between quenching and recharging of the SPAD voltage.
- **8-Bit Ripple Counter**: The photon integrator is implemented by means of an 8-bit ripple counter. This counter performs the vital function of quantifying the photon counts detected by the SPAD. Its static flip-flop configuration ensures reliability, particularly when employed as a memory element. While alternative integration techniques such as analog counters or continuous-time integrators are viable [Dutt16; Panc13; Pere16], the decision to implement an 8-bit digital counter was made with consideration for its role as an in-pixel memory, favoring stability over volatile alternatives.
- **Multiplexer for Event Detection**: One of the pivotal features of the digital counter is its capacity to facilitate event detection based on selected output bits. This selection effectively establishes a relationship between an event and the arrival of N_{ph} photons, with N being defined by the chosen counter output bit.
- **4-Bit SRAM Memory**: Inclusion of a 4-bit SRAM module serves to manage the state of the pixel, enabling both enable/disable functionality and selection of the counter output bit. This memory unit, though only operating in write mode, offers a valuable avenue for on-line adjustment of N_{ph} sensitivity.
- **AER Readout Logic**: At the final stage, the AER readout logic takes charge, employing pull-down transistors to manage shared signals at both column and row levels. This integral logic component seamlessly implements the AER protocol, facilitating the efficient transmission of events beyond the pixel array.

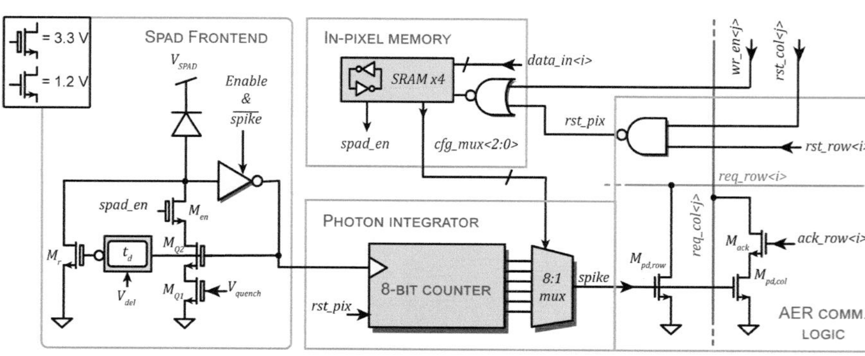

FIGURE 4.2 Pixel schematics. The pixel is composed of a SPAD front-end, a multi-purpose 8-bit counter, a 4-bit SRAM memory, and AER logic.

It is important to note that while this specific implementation is presented, alternative approaches for the different blocks illustrated in Figure 4.1 are viable and worth exploring. Subsequent sections will provide an analysis of each element mentioned earlier, discussing alternative implementations and the design choices associated with them.

SINGLE-PHOTON RECEPTOR

At the heart of the functionality of the pixel is the SPAD, responsible for the precise detection of individual photons. This subsection delves into the quenching and recharge circuitry of the SPAD that, as explained in Chapter 1, are critical processes governing the termination of the avalanche effect and rapid restoration to a quiescent state [More19; Pere18; Seit13].

Figure 4.3 provides detailed perspectives of the SPAD device. The junction employed for the device is a p-well/deep n-well structure with a poly guard ring surrounding the device to avoid field-gate oxide formation close to the device [Vorn21]. This is essential to avoid premature-edge breakdown and reduce impurities close to active area. Also, note that the *Deep N-Well* (DNW) available in this technology is a retrograde DNW. This is particularly interesting because it provides an effective means to confine the high electric field in the desired region (multiplication region) and prevent premature-edge breakdown [Vorn21].

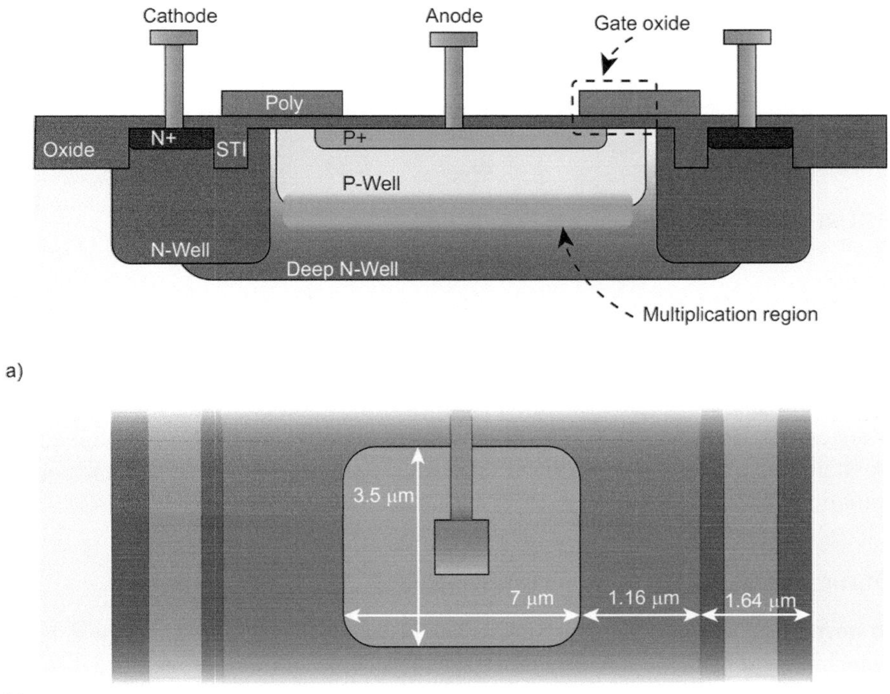

FIGURE 4.3 a) SPAD cross-section. b) SPAD top view.

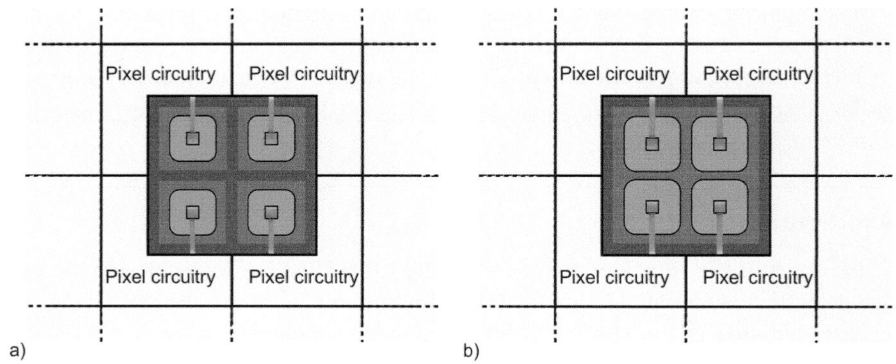

FIGURE 4.4 a) Current arrangement of SPADs, sharing DNW among 2×2 pixels. b) Further Fill Factor improvement by sharing guard ring.

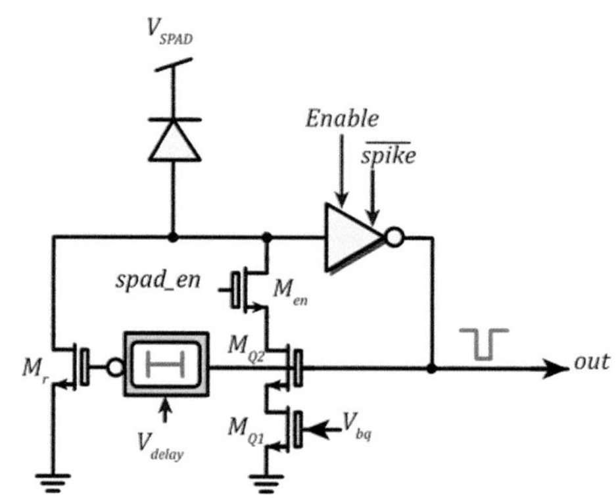

FIGURE 4.5 SPAD front-end of the SAER pixel.

Finally, SPADs are arranged in groups of 2×2 pixels sharing the DNW, as depicted in Figure 4.6 (a). This allows increasing fill factor since DNW design rules are usually very restrictive in terms of design rules. Further improvement may be achieved by sharing the guard ring among pixels [Mori20b]. Figure 4.6 (b) depicts this arrangement. However, SPAD devices in close proximity are likely to exhibit a high crosstalk probability. Thus, pixels of the SAER vision sensor implement the configuration shown in Figure 4.6 (a).

Quenching-and-recharge circuitry

To operate such a junction as a SPAD, it is necessary to include the corresponding quenching and recharge circuitry. This circuitry, depicted in Figure 4.5 is a detail of Figure 4.2 showing the circuitry employed for quenching and recharge. It implements an active-quench and active-recharge mechanism. It is composed of transistors M_{Q1}, working as a tunable resistor, and M_{Q2}, which operates as a non-linear resistor. The incremental resistance of M_{Q1} is determined by the voltage V_{bq}, and expressed as:

$$R_{Mq1} \simeq \frac{L}{\mu_n C_{ox} W \left(V_{qs} - V_{tn}\right)} \quad (4.1)$$

while the resistance of M_{q2} increases with current. This behavior is given by a positive feedback loop involving the output inverter. As the drain voltage of M_{q2} increases, the gate voltage decreases, increasing the resistance of the device. This structure is similar to that reported in [Vorn16], but the presence of M_{q1} adds a second positive feedback path, since the source voltage of M_{q2} also increases with the current.

To complete the quenching and recharge process, the output of the inverter is negated and delayed before driving transistor M_r, which executes the active recharge process. This dynamic interaction ensures efficient quenching and timely restoration of the SPAD for subsequent photon detection events.

Figure 4.6 complements this circuitry by illustrating the implementation of the delay element. This element is realized through a current-starved inverter, where the delay is precisely defined by the ratio of the driving current to the output capacitance. The driving current can be finely controlled by adjusting the voltage V_{delay}, providing a mechanism to fine-tune the dead time of the SPAD. This voltage is in fact generated by programming a reference current.

Photon integrator and comparator

The core functionality of the pixel depends on the photon integrator and comparator elements, each playing a crucial role in signal processing. As explained in Section 4.1.2, the design choice was implementing a digital counter that implements both the photon integrator and the comparator function. A key advantage of the digital counter is its capacity to facilitate event detection based on selected output bits. This selection effectively establishes a relationship between an event and the arrival of N_{ph}

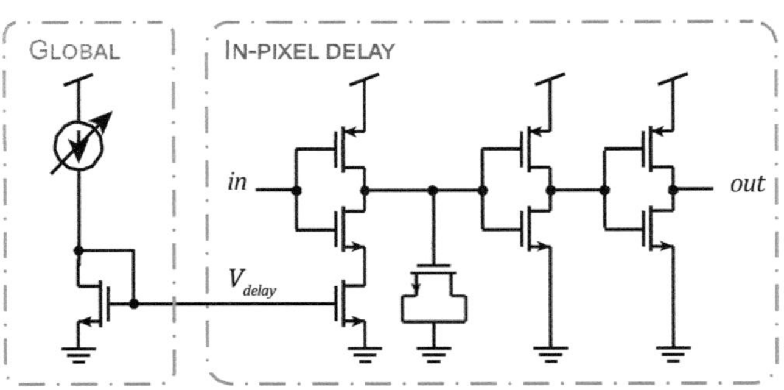

FIGURE 4.6 Delay element implementing the dead time of the SPAD.

photons. This characteristic enables precise control over event generation, allowing for adaptive response to varying photon arrival rates.

The choice of N_{ph} is a critical parameter in the operation of the pixel. It fundamentally governs both the frequency of the output signal and the statistical properties of event generation. Since photon arrival follows a Poisson process, the inter-arrival time between photon events, or the period of the output signal, conforms to an exponential distribution. This property enables a priori knowledge of the SNR and the coefficient of variance of the distribution, equating to $\sqrt{N}$ and $1/\sqrt{N}$, respectively. This knowledge empowers precise control over the statistical properties of the output pixel, offering valuable insights for subsequent processing stages.

Figure 4.7 (a) depicts the ripple counter, a series of cascaded D flip-flops. Figure 4.7 (b) offers insight into the implemented flip-flop, employing a master-slave configuration of inverter-based latches and transmission gates. Notably, one of the latches incorporates a NAND gate for reset operation.

An alternative to save space, allowing for an increased counter bit count or lower pixel pitch, is the use of dynamic flip-flops [Tang18]. Figure 4.8 shows the simplest implementation of a dynamic flip-flop, which also operates in a master-slave fashion, but instead of using latches as memory elements, the signal is stored in capacitive nodes.

Although V_x and Q store the same signal, the output inverter is required if the flip-flop is connected in feedback configuration. This architecture reduces area and timing limitations, at the cost of leaky storage nodes. However, we opted for static flip-flops. This decision was motivated by the extra functionality that static flip-flops provide, allowing the counter to serve as a memory for extended periods of time.

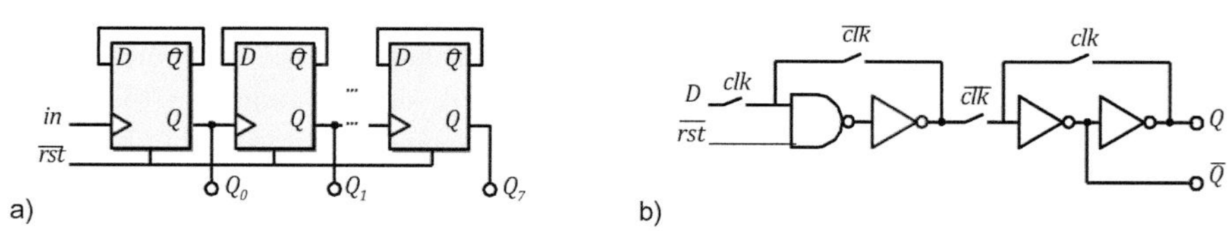

FIGURE 4.7 a) Concept of ripple counter. b) Master-slave D flip-flop implementation.

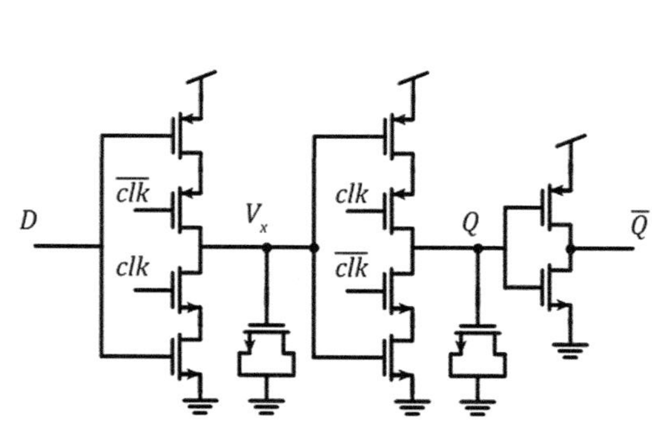

FIGURE 4.8 Implementation of a dynamic flip-flop based on C2MOS logic.

Also, the capacitance of nodes V_x and Q can be adjusted for the most significant bits to prevent data loss due to leakage, since data is inherently stored for a longer time than in less significant bits.

Analog alternative for the photon integrator

Transitioning to alternative implementations, the analog domain offers a feasible path for implementing the photon integrator. Figure 4.9 provides an example of an analog counter based on a charge transfer amplifier [Dutt16]. In this implementation, the voltage step corresponding to an input pulse, ΔV, can be expressed as [Dutt16]:

$$\Delta V = \frac{C_o}{C_p}\left(V_{IN} - V_{tn1} - V_s\right) \tag{4.2}$$

where V_{IN} is the amplitude of the input pulse, V_{tn1} is the threshold voltage of the input transistor, M_{n1}, and V_S is a constant voltage. Therefore, equation (4.2) shows how the voltage can be tuned by adjusting the value of V_S, keeping in mind that V_G must be high enough to keep M_{n2} in the saturation region. Also, M_{p1} can be substituted by a NMOS transistor in order to have a NMOS-only circuit [Dutt16].

Analog counters have demonstrated excellent performance, exhibiting an acceptable level of mismatch [Dutt16; Panc13; Park21]. However, it is important to acknowledge the trade-offs. While they excel in area efficiency, analog counters introduce electronic noise, both from the counter itself and the subsequent analog comparator that would be required to generate the events.

Furthermore, the in-pixel digital integrator can serve as a memory element, retaining photon counts between exposures with no leakage. This capability results in a special interest in operating modes such as QI or ToF where pre-integrated binary frames reduce the number of events that are transmitted outside the pixel. Therefore, the choice for a digital counter

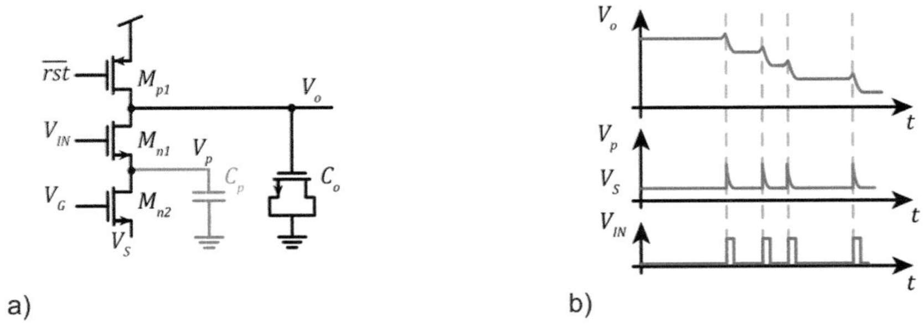

FIGURE 4.9 a) Schematics of an analog counter. b) Waveforms when operating as a Charge Transfer Amplifier.

AER READOUT LOGIC

The method of transmitting events from a pixel upon their generation involves diverse readout circuitry designs, varying based on specific requirements and specifications of the sensor. There are two primary transmission mechanisms: asynchronous transmission and scanning mechanisms. Determining the optimal solution involves considering several factors, including the pixel count, event characteristics, and expected activity.

Utilizing an asynchronous readout scheme presents a significant downside: the unpredictable sequence of event handling leads to timing errors, especially among neighboring columns simultaneously generating events [Suh20]. This discrepancy becomes particularly pronounced in sensors with high spatial resolution. Consequently, many sensors with large resolutions have transitioned to scanning schemes to manage events more effectively [Guo23a; Koda23].

However, asynchronous communication results in a more efficient utilization of the bandwidth in the case of low-resolution prototypes. The fewer elements accessing to the readout channel has a lesser corrupting impact on temporal signals. Therefore, the SAER pixel includes an asynchronous AER readout logic to transmit events outside the pixel array.

The AER readout logic is the element that guarantees efficient data transmission from the pixel to the peripheral circuitry. At the heart of the AER protocol lies a network of pull-down transistors, distributed along the pixel array, that control shared signals at both the column and row levels. This arrangement facilitates precise signal routing, allowing for event acknowledgment and subsequent data transmission.

Upon the pixel generating an event, the AER readout logic is in charge of generating the signals that implement the AER protocol, as shown in Figure 4.10. First, $M_{pd,w}$ pulls down the common signal req_row. This action initiates the arbitration process at the row level, eventually enabling ack_row, which allows $M_{pd,l}$ to repeat the request process at the column level.

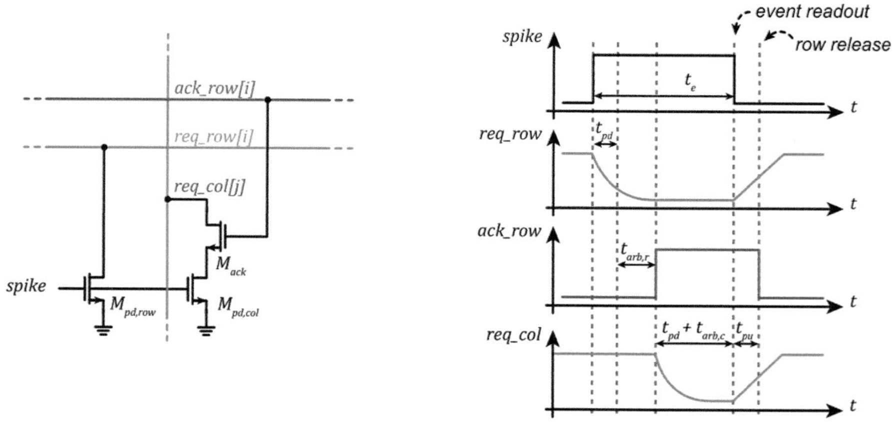

FIGURE 4.10 AER logic in the SAER pixel and waveform diagram.

When the external receiver successfully receives the event, it triggers the activation of rst_row and rst_col signals, which have been omitted in Figure 4.10. These signals, in turn, generate the internal reset signal, rst_pix initiating the reset of the counter and preparing the pixel for subsequent event detection.

Several delays are inherent in this process, making it imperative to understand and characterize these delays, particularly considering that event information is encoded in the time domain. Figure 4.10 illustrates various types of errors, including:

- $\boldsymbol{t_{pd}}$: Pull-down time from event generation to activation of req_row. It depends on the on resistance of pulldown transistors and line impedance.
- $\boldsymbol{t_{arb}}$: Arbitration time, the period from request reception to acknowledge generation. It depends on AER readout block and activity.
- $\boldsymbol{t_{pu}}$: Pull-up time. Influenced by the strength of the pull-up transistor and line impedance.
- $\boldsymbol{t_{e}}$: Total latency from event generation to event readout.

While t_{pd} and t_{pu} tend to remain constant under ideal conditions, t_{arb} exhibits a significant random factor due to its dependency on the number of events managed by the AER readout block. Its correlation with t_{pu} is substantial; as the latter increases, it enlarges the time required to read pending events before permitting access to the specific pixel. Furthermore, t_e diminishes when reading pixels from the same row, as these pixels circumvent the need for arbitration time at the row level since access is already granted. Thus, taking all these facts into consideration and employing techniques mentioned in Section 3.4 are crucial to reduce t_e, and hence, signal corruption.

4-Bit SRAM memory

The 4-bit SRAM memory serves as a crucial element in governing the state of the pixel and controlling its output. While Figure 4.11 (a) shows the schematic of a conventional 6T-SRAM cell [Sing13], the memory element implemented in the SAER pixel is shown in Figure 4.11 (b).

Conventional SRAM cells are managed by a word line, *WL*, which links its output, Q, and its complement, *Q*, to a bit line, *BL*, and its complement, *BL*. This arrangement is necessary for the writing and reading operation to either set or create an imbalance between both signals, respectively. Nonetheless, this setup demands two wires per

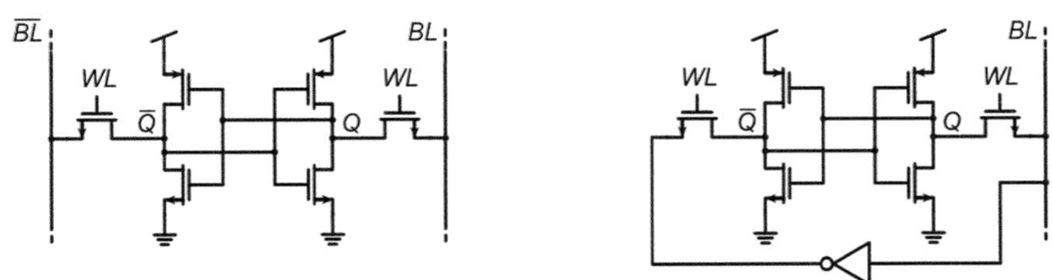

a) b)

FIGURE 4.11 a) Conventional 6T-SRAM cell. b) SRAM cell of the SAER pixel.

bit, posing a constraint within a pixel. Doubling the number of wires constrains fill factor and pixel pitch.

In a departure from the conventional 6T-SRAM design, this SRAM cell operates with only positive data input. The negative data signal is generated within the cell through the integration of an additional inverter. The reason for selecting this unconventional approach is twofold. Firstly, it is not necessary to read the information stored in the SRAM memory, allowing a write-only operation. Secondly, it eliminates the need for incorporating four extra wires, a design choice made to optimize area consumption.

To further optimize space utilization, data inputs are shared across every two rows, as illustrated in Figure 4.12. This decision minimizes congestion within the pixel array, at the cost of defining different phases to program odd and even rows. Note that the impact in layout is minimal, since only difference between pixels from odd an even rows is the position of metal vias, minimizing any non-uniformity due to non-identical layout. By carefully balancing functionality with resource allocation, this design approach ensures a proper area efficiency.

While the custom SRAM represents a notable reduction of wiring, it has also some drawbacks as explained below. A critical consideration arises from the use of NMOS transistors as switches for data input. In this configuration, as in the conventional SRAM, logic 0 (be it the negated or positive data input) is the driving force for data writing. However, due to the exclusive use of positive data input, writing logic 1 takes slightly longer. This leads to a brief period during which the SRAM draws static current. While seemingly subtle, this timing discrepancy can have profound implications, especially when attempting to write the entire pixel array simultaneously. In such a scenario, the collective demand for current can precipitate a noticeable voltage drop in both the power supply and ground lines. This, in turn, may introduce metastability issues in the memories, potentially compromising data integrity. To mitigate this, as explained in Section 3.3.2, a thoughtful strategy involving a staggered generation of the rst_col is employed. This deliberate shift in timing ensures that the memory of the last column is updated prior to the beginning of the writing process for the subsequent column, safeguarding against potential current peaks in the SRAM memory.

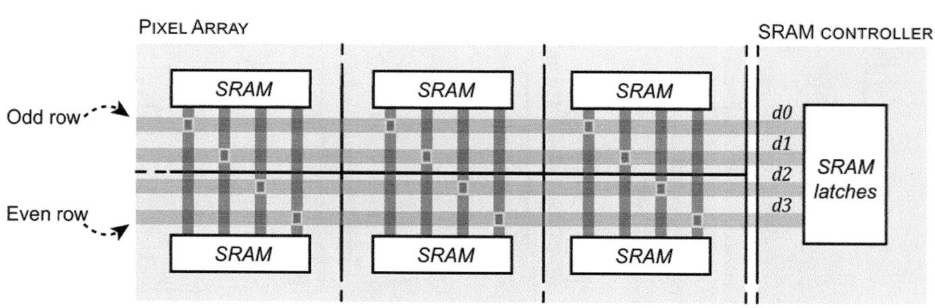

FIGURE 4.12 SRAM input interconnection shared between adjacent rows.

4.2 EXTENDING THE DYNAMIC RANGE

The pixel architecture employed in our sensor not only presents the capability of working in single-photon-detection mode but also accommodates a valuable alternative mode: operating as a conventional linear-mode photodiode. This versatility may be particularly useful in scenarios where conventional SPAD devices might face saturation or when the output frequency of the pixel reaches levels that saturate the bandwidth of the readout channel. By biasing the photodiodes below the avalanche voltage, the sensor can effectively adapt to high-intensity scenes, thus expanding its dynamic range and ensuring proper operation in a wide array of scenarios.

4.2.1 PIXEL CONFIGURATION IN THE I&F MODE

When the photodiode is biased in linear mode, the pixel can work as an *Integrate and Fire* (I&F) neuron. This time the photodiode is the one that integrates photogenerated carriers, and the 8-bit counter can be used to divide the frequency of the output pulses.

In the I&F mode, the quenching and recharge circuitry is reconfigured to function as a self-reset mechanism. This modification entails a significant departure from its primary single-photon detection mode. Figure 4.13 (a) provides a visual representation of the equivalent circuit, offering insight into the intricacies of this modified configuration, while Figure 4.13 (b) represents the corresponding waveform diagram when N_{ph} is selected to 4. In this mode, the in-pixel counter can be used to further adapt the sensitivity of the pixel if the bandwidth of the readout channel saturates.

This modification is accomplished by setting V_{bq} to 0 V, thus disabling the quenching branch and biasing the photodiode below the avalanche region, e.g., setting V_{SPAD} to 3.3 V. Then, the output inverter acts as the comparator that determines that the photodiode collected a certain number of photons. Therefore *spike* is enabled after collecting N', which can be expressed as:

$$N'_{pn} = N_{pn} \frac{C_{SPAD} \cdot V_{th}}{q} \tag{4.3}$$

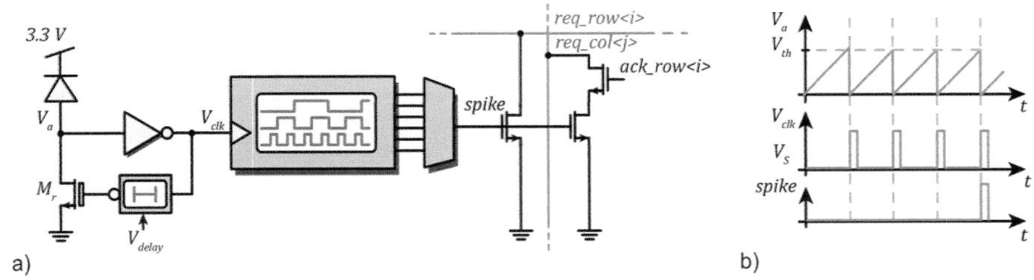

FIGURE 4.13 a) Pixel configuration in the I&F mode. b) Corresponding waveforms.

where C_{SPAD} is the equivalent capacitance at the anode of the diode and V_{th} is the threshold voltage of the inverter. Then, after the output of the inverter is enabled, the recharge transistor, M_r, resets the anode to ground, starting the photon collection operation once again.

This reconfigured pixel mode, tailored for I&F functionality, enables an extension of the dynamic range, adapting to diverse operational scenarios. However, it is not trivial to optimize the sensor for both single-photon and I&F mode. Limitations regarding the comparator or the reset mechanism appear as a trade-off when designing such components.

4.2.2 Design considerations for the I&F mode

Optimizing the sensor for I&F functionality entails a distinct set of considerations compared to its single-photon mode. Each design choice should be made with the explicit goal of ensuring proper operation in this alternative mode; however, these considerations may affect the efficiency of the sensor when working in single-photon detection mode. One critical aspect lies in the dependence of the photon threshold of the I&F, denoted as N_{ph}', on V_q, as expressed by equation (4.3). This parameter is defined by the aspect ratio of the PMOS and NMOS transistors in the CMOS inverter.

As we seek to achieve precise matching, it becomes evident that using larger devices is necessary. However, this design choice introduces a trade-off; while it effectively minimizes mismatches, it concurrently increases the input capacitance. Consequently, the anode signal experiences extended rise and fall times during single-photon mode operation, which is undesirable. Besides, as the threshold depends on both PMOS and NMOS devices, no accurate relationship can be established.

In tandem with this consideration, optimizing the output buffer for speed and power efficiency becomes essential. The implementation of a CMOS inverter instead of an alternative architecture arises from the goal of optimizing these two metrics, due to its inherently class AB nature. Nevertheless, when the input to this stage takes the form of a ramp rather than a sharp pulse, there arises the potential for a large current consumption, particularly when the anode voltage approaches V_q. While the implementation of a common-source amplifier could mitigate this issue, this comes at the cost of requiring an additional bias voltage, potentially compromising the area efficiency and routing congestion of the pixel.

Moreover, properly sizing the quenching and recharge branch is crucial. In single-photon mode, the ideal scenario entails a robust transistor capable of rapidly recharging the SPAD voltage, while avoiding self-loading. However, pushing the strength of the device to its limit could increase leakage current during the off-state, thereby introducing pixel-to-pixel non-uniformity when it operates in linear-mode.

Implementing a hold-off time in addition to the hold-on time, which is governed by the delay in the quenching and recharge circuit, would also be advantageous for ensuring a proper reset of devices. However, this requires the introduction of an additional bias voltage, potentially impacting the area performance of the pixel.

These considerations serve as guiding principles throughout the optimization process of the I&F mode, along with the trade-offs between single-photon detection and I&F mode, each with its unique set of operational requirements. It is crucial to acknowledge that these design trade-offs have led to the optimization of the SAER pixel primarily for single-photon mode operation. Consequently, while I&F mode is supported, it is not anticipated to deliver the optimal performance, since the single-photon functionality was prioritized.

4.3 ENHANCING THE FUNCTIONALITIES OF THE SAER PIXEL

The SAER pixel, as described thus far, provides a robust architecture for event-based imaging capabilities using SPAD devices. However, the potential for further enhancement and expansion of functionalities is vast. This can be achieved through a combination of well-established techniques commonly implemented in SPAD-based sensors, as well as leveraging the potential offered by technological advancements such as 3D stacking. By exploring these avenues, we can unlock a range of possibilities for refining pixel performance and incorporating innovative circuit architectures, ultimately pushing the boundaries of imaging capabilities.

Note that the additional functionalities outlined in this section were not incorporated into the experimental prototype. This was primarily due to constraints in the available area, which limited our ability to achieve a reasonable pixel resolution. Nonetheless, it is worth highlighting that other SPAD-based architectures, as proposed by various researchers, have illustrated the potential for integrating advanced digital processing resources at the pixel level [Gyon23; Rocc20] or in a logic die [Hutc19]. This can be accomplished when utilizing a smaller node or a larger pixel pitch, as demonstrated in the referenced literature.

4.3.1 MULTIPLE SPADs PER PIXEL

As depicted in Figure 4.14, when a SPAD operates under high photon flux, a notable number of incident photons are lost due to the dead time of the device. This crucial characteristic imposes a restriction on the maximum achievable frequency of the SPAD, with the choice of dead time typically falling within the range of tens to hundreds of nanoseconds, contingent upon the specific device.

However, it is imperative to recognize that limitations arise not only in scenarios of excessive illumination. The same effect occurs when the SPAD is exposed to a

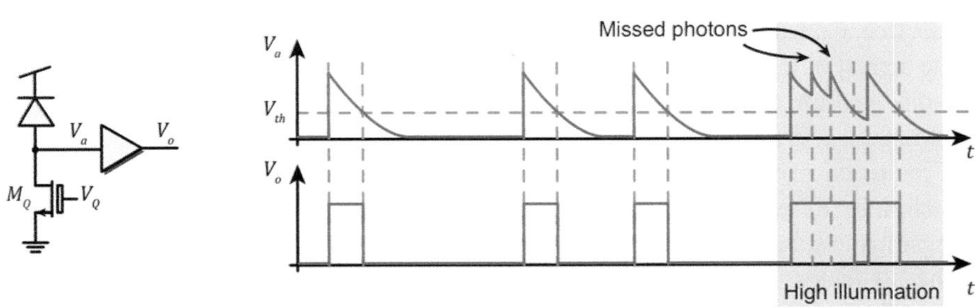

FIGURE 4.14 Illustration depicting lost photons due to a high photon flux. It is assumed that the SPAD employs passive quenching, with MQ operating in the linear region.

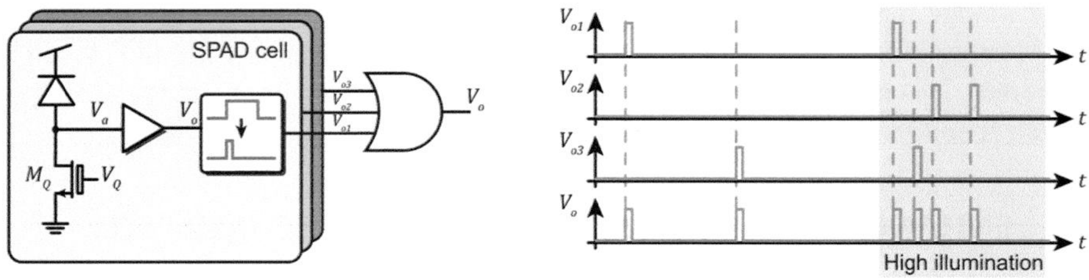

FIGURE 4.15 Concept of multiple SPADs per pixel. The output of the SPADs is shortened and ORed.

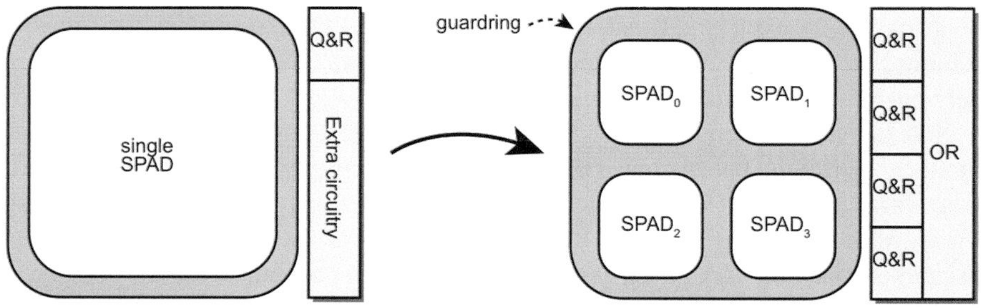

FIGURE 4.16 Transition from a pixel implementing a single SPAD to a macro-pixel including four SPADs.

correlated light source, such as a laser pulse. Given that the duration of the laser pulse is often in the range of several nanoseconds, a SPAD detector is inherently constrained in its ability to register more than one photon during each exposure. This limitation is translated to the performance of the pixel, requiring multiple laser exposures to achieve a given SNR. Consequently, this presents an influence on both the frame rate and overall power consumption of the system.

To overcome this limitation, one viable solution consists of incorporating multiple SPAD devices within each pixel [Vorn20], as illustrated in Figure 4.15. Each SPAD, equipped with its corresponding quenching and recharge circuitry, operates autonomously. The outputs of these individual SPADs are then shortened and connected to an OR tree. Consequently, the output signal from the OR tree is a pulse corresponding to the arrival of a photon at any of the SPADs within the pixel, effectively improving photon detection capabilities.

However, it is important to note that this approach introduces a notable trade-off. The inclusion of multiple SPAD devices within a single pixel inevitably leads to a reduction in fill factor. This occurs because of the need for additional guard ring area, alongside the supplementary circuitry, which collectively contributes to the reduction of the fill factor. Figure 4.16 illustrates the transition from a pixel implementing a single SPAD to a macro-pixel that includes four SPADs.

4.3.2 Photon-coincidence for Time-of-Flight

In complement to incorporating multiple SPADs within pixels, implementing coincidence detection mechanisms at the pixel level enhances pixel robustness against background illumination [Nicl13; Padm21]. Background light usually comprises uncorrelated photons governed by Poisson statistics. Conversely, photons originating from coherent light sources like lasers are correlated and expected to arrive within close proximity in time. Thus, when a pixel contains more than one SPAD, photon detections can be filtered based on the number of SPADs detecting a photon concurrently.

This filtering process lowers the probability of detecting uncorrelated photons within a specific time interval. As the number of SPADs required to detect a coincidence, N_c, increases, the probability of coincidence decreases. This probability, P_c, can be mathematically defined as:

$$P_c = \prod_{i=1}^{N_c} P_i \tag{4.4}$$

Here, P_i represents the probability of SPAD i to detect a photon within time T, assuming uncorrelated photons arriving at each detector. When the distance between SPADs is relatively small, photon flux of all detectors can be assumed similar and equal to ϕ. In this scenario, equation (4.4) can be rewritten as:

$$P_c = \left(1 - e^{-\phi T}\right)^c \tag{4.5}$$

Different methodologies exist for performing this detection at the pixel level. For instance, Figure 4.17 (a) shows a circuit employing a sub-counter to determine photon detection when 2^n photons arrive within the time interval defined by *enable*. However, this circuit requires the dead time of SPADs to be longer than T in order to avoid multiple counts from the same SPAD. Also, this approach may filter pulses if photons coincide during the pulse of other SPADs, resembling no coincidence, as shown in Figure 4.17 (b).

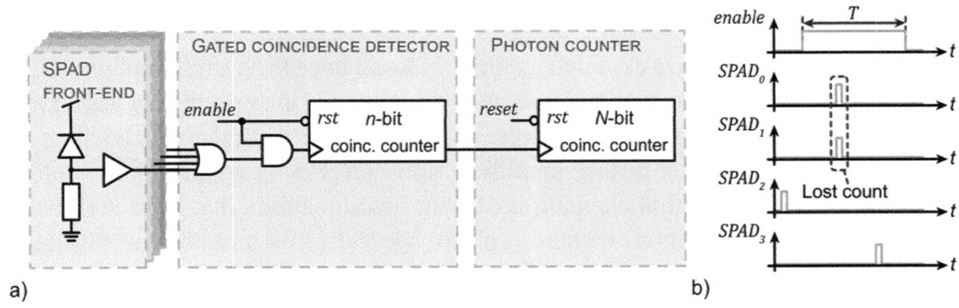

FIGURE 4.17 a) Counter-based coincidence detector. b) Lost counts if detection occurs during the pulse width of a SPAD pulse.

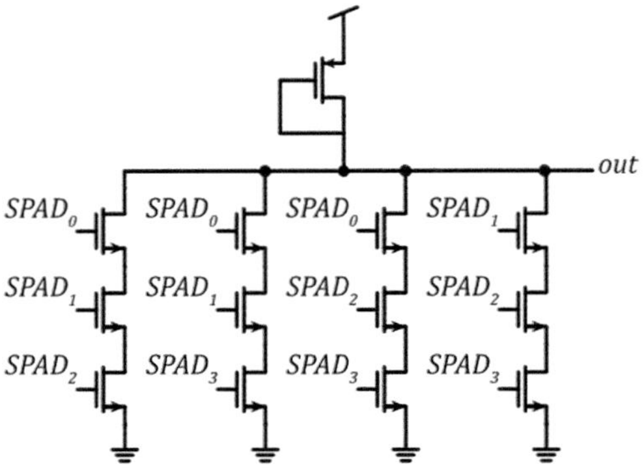

FIGURE 4.18 Example of 3-out-of-4 coincidence detector based on logic gates.

To address both limitations of the circuit in Figure 4.17 (a), a coincidence detector based on logic gates can be implemented. Figure 4.18 shows a 3-out-of-4 coincidence detector implemented as a logic gate, where a diode-connected PMOS was employed to trade between area and power consumption. Also, if the sensor was not intended to count photons, the OR tree can be substituted by the circuit. However, if fewer coincidences were to be used to trigger a detection, more combinations would appear, increasing the number of transistors and routing complexity. The operation can be also implemented in the analog domain [Nicl10]. However, these techniques are prone to mismatch and more power-hungry.

In practical application, photon coincidence faces limitations. Detectors might not register all photons due to limited PDP, reducing the SNR of the resulting histogram. Furthermore, counting photons based on coincidences negates the benefits described in the previous section where a pixel can detect multiple photons during a single exposure, reducing even more the SNR. To do so, additional logic to both count detections and add this count to a bin counter when coincidence occurs is required [Nicl13], making it unfeasible at the pixel level. The complexity involved in implementing this technique at the pixel level has restricted its widespread adoption in ToF detectors.

4.4 ADDITIONAL SAER PIXEL'S FUNCTIONALITIES

In the preceding section, we explored techniques to enhance the current functionalities of the SAER pixel, focusing on refining its performance using established techniques provided that sufficient area is available. However, the potential for extending the capabilities of the pixel is further amplified when leveraging advanced technology nodes. This entails the implementation of new functionalities that hold significant promise from an application-oriented point of view. In this section, we will delve into the adaptation of the pixel architecture to facilitate the integration of temporal contrast-detection capabilities, broadening the scope of potential applications in a future implementation.

4.4.1 In-pixel temporal contrast detection

Events within the SAER vision sensor can be classified into intensity and motion events, as described in Section 2.3.3. However, the event classifier is implemented in the CPM, potentially causing a bottleneck at the AER readout block. This bottleneck is related to continuous event transmission, even when the scene remains static.

To alleviate this bottleneck, it would be advantageous to implement the event classifier within the focal plane. This implementation could reside inside individual pixels, within clusters, or on a logic die leveraging the high density of advanced technology nodes. By integrating the event classifier into the sensing devices, it could mitigate readout channel bottlenecks, remain immune to event collisions, and simplify the overall implementation process.

Digital discrete DVS

Figure 4.19 depicts a possible implementation of an in-pixel event classifier. The concept is the same as in Section 2.5.3; however, the only memory element required in this implementation is the reset register that stores the last inter-arrival time that triggered an event, T_{rst}. The goal of the operation is to detect the condition when $|T_k - T_{rst}| > \alpha_{th} T_k$, where T_k is the current inter-arrival time and α_{th} is the contrast threshold. The operation is as follows:

1. The time counter is continuously counting clock cycles until the SAER pixel generates a *spike* to measure the current inter-arrival time, T_k.
2. Then, the event classifier computes $|T_k - T_{rst}|$.
3. The event classifier compares $|T_k - T_{rst}|$ to the LSBs of T_k.
4. After a defined delay, t_d, to account for propagation delay, the output of the comparator and sign are registered. If a temporal event was detected, T_{rst} is updated.
5. Another delay is implemented to ensure proper update of T_{rst} before generating the signal *rst_pix* that resets the SAER pixel.

While the shift operation of T_k is simplified to rearranging its wires, the primary limitation of this circuit is encountered in implementing the subtractor, which requires

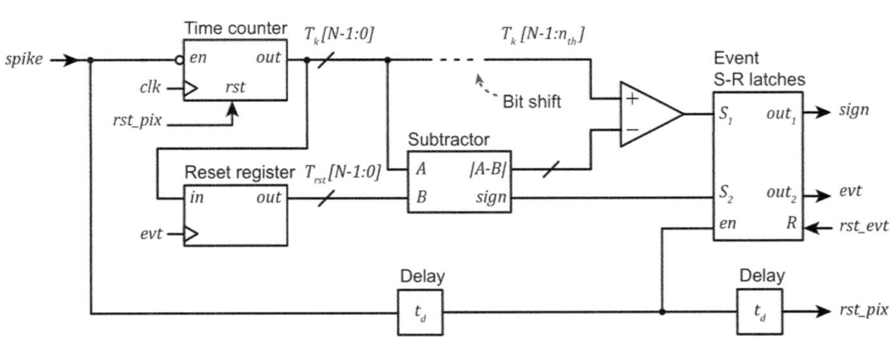

FIGURE 4.19 Example of implementation of an in-pixel event classifier.

a greater number of logic gates. Additionally, it is worth noting that the comparator can be simplified, as the MSBs of T_k are compared to 0. As for the initial state, an additional register indicating that the reset register is empty must be included to avoid wrong comparisons at the beginning of the operation.

Therefore, this solution remains suitable for sensors leveraging 3D stacking processes [Guo23a; Wuu22], where event processing may be performed at the cluster level. Thus, the subtractor can be shared among different pixels and accessed after spike generation. However, this also implies arbitrating the access of the shared resource among pixels.

Analog discrete DVS

The discrete DVS operation can also be implemented in the analog domain. To achieve functionality akin to the original approach [Lich08], the circuit should encompass the following components:

- A ***Time-to-Voltage Converter*** (TVC) with an ideal logarithmic response to time, mimicking the behavior of logarithmic photoreceptors to inherently measure relative variations. Consequently, variations in this block between pixels do not impact the operation.
- A **sampling circuit** that captures the TVC output each time a spike occurs.
- A **memory element** to retain the TVC output from the previous event.
- An amplifier capable of amplifying variations in the sampled signal relative to that of the memory element.

The primary constraint arises during the implementation of the TVC. The simplest method involves a current source charging a capacitor, resulting in a linear conversion between time and voltage. A circuit demonstrating a logarithmic relationship is depicted in Figure 4.20 (a). In this setup, a MOS transistor, M_{log}, operates in the sub-threshold region. Consequently, if the gate voltage approximates threshold voltage of the transistor, V_{th}, and we neglect body effect and channel length modulation, the current flowing through the capacitor, I_c, can be expressed as:

$$I_c \simeq I_o e^{\frac{V_b - \frac{V_{\log}}{nU_t} - V_{th}}{}} = I_o e^{-\frac{V_{\log} - \Delta V}{nU_t}} = C_t \frac{dV_{log}}{dt} \tag{4.6}$$

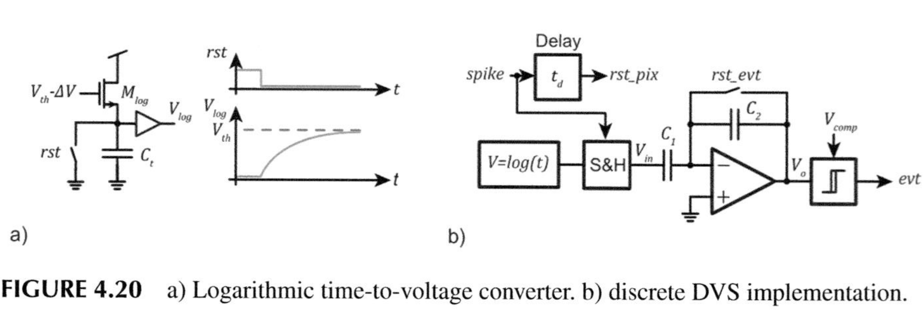

FIGURE 4.20 a) Logarithmic time-to-voltage converter. b) discrete DVS implementation.

Solving the differential equation for V_{log} results in the following expression:

$$V_{log}(t) = \left| \frac{I_c}{C_t} dt \simeq nU_t \text{In} \right| \left(\frac{I_o}{nU_t C_t} e^{\frac{-\Delta V}{nU_t}} \right) t + 1 \tag{4.7}$$

where n is the sub-threshold slope factor and I_o is the scale factor of the current depending on technology parameters and the size of the device [Enz06]. Note that the bias voltage can be generated from a diode-connected device biased with a small current to ensure the circuit is biased in the sub-threshold region regardless of PVT variations.

Equation (4.7) indicates that there is a lower limit for the time at which the circuit functions effectively. This specifically refers to the point where the circuit demonstrates a logarithmic dependence on time, i.e., when the term relying on t becomes dominant compared to the term dependent on ΔV, i.e., when $t \gg \frac{nU_t C_t}{I_o} \cdot \frac{\Delta V}{e^{nU_t}}$ point, V_{log} can be approximated as:

$$V_{log}(t) \simeq nU_t \text{In}(t) + nU_t \text{In}\left(\frac{I_o}{nU_t C_t} \right) - \Delta V \tag{4.8}$$

However, while the circuit is expected to ideally reach V_{th} and even higher voltages with prolonged operation, an upper limit on V_{log} is set by the parasitic diode located at the source of M_{log}. This limitation occurs when the current flowing through M_{log} equals the leakage current of that diode. Consequently, to detect a broader range of frequencies, it is imperative to minimize both ΔV and the source area of M_{log}.

The TVC can be then included in the circuit of Figure 4.20 (b). In this configuration, C_1 and C_2 act as memory elements akin to conventional DVS, and the output of the TVC is sampled after the rising edge of *spike*. Hence, by defining V_k as the voltage sampled upon the rising edge of *spike* and V_{rst} as the value of V_k when *rst_evt* was triggered, the circuit generates events when the following condition is met:

$$\left| V_k - V_{rst} \right| \frac{C_1}{C_2} > V_{comp} \tag{4.9}$$

Assuming that the TVC operates in the logarithmic region, and defining temporal contrast as $\alpha = T_{rst}/T_k - 1$, equation (4.9) can be rewritten as:

$$\left| nU_t \ln\left(\frac{T_k}{T_{rst}} \right) \right| \frac{C_1}{C_2} > V_{comp} \tag{4.10}$$

$$\left| nU_t \ln(1+\alpha) \right| \frac{C_1}{C_2} > V_{comp} \tag{4.11}$$

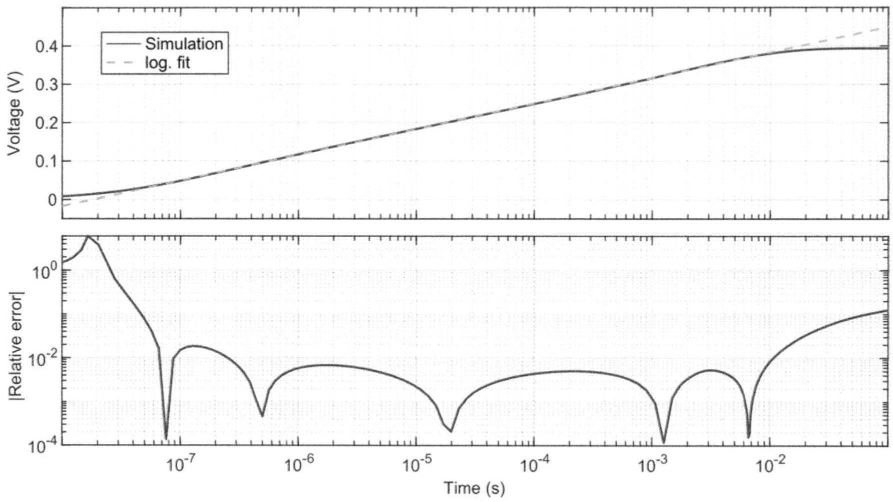

FIGURE 4.21 Simulation results of the logarithmic Time-to-Voltage Converter.

Therefore, α_{th} can be defined as the value of α that satisfies equation (4.11), resulting:

$$\alpha_{th} = e^{\frac{|V_{th}|C_2}{nU_tC_1}} - 1 \tag{4.12}$$

Equation (4.12) also suggests that the absolute value of V_{comp} must increase with temperature to ensure a fixed contrast threshold.

Figure 4.21 illustrates the simulation results of the TVC and a logarithmic fit. Upon reaching the region where the term dependent on t dominates, the error drops below 1%. Notably, in this design, the TVC exhibits a dynamic range of 100 dB. Equations (4.8) and (4.12) demonstrate that parameter mismatch in the TVC does not affect its operation. Mismatch primarily impacts the operational range. Consequently, the circuit demands that C_t remains linear within the operation range, and the I_o/C_t ratio must be carefully selected to match the operational range of the SAER pixel. In practice, the amplifier from Figure 4.20 (b) is implemented using a single ended architecture, leading to variations in the effective value of V_{th} from pixel to pixel.

Figure 4.22 depicts the simulation results of the circuit using ideal components, excluding the TVC. The circuit operates similarly to a conventional DVS, but now in a discrete manner. Additionally, the signal reconstructed from events aligns well with the input frequency. Nevertheless, the reconstructed signal is slightly distorted due to the discrete operation.

While continuous-time DVS can accurately trigger events at the precise moment when temporal contrast occurs, discrete DVS may detect contrast thresholds higher α_{th}, depicted in Figure 4.23, especially when dealing with low intensity and rapid transients. This effect is noticeable in Figure 4.22, when V_{out} over-exceeds the

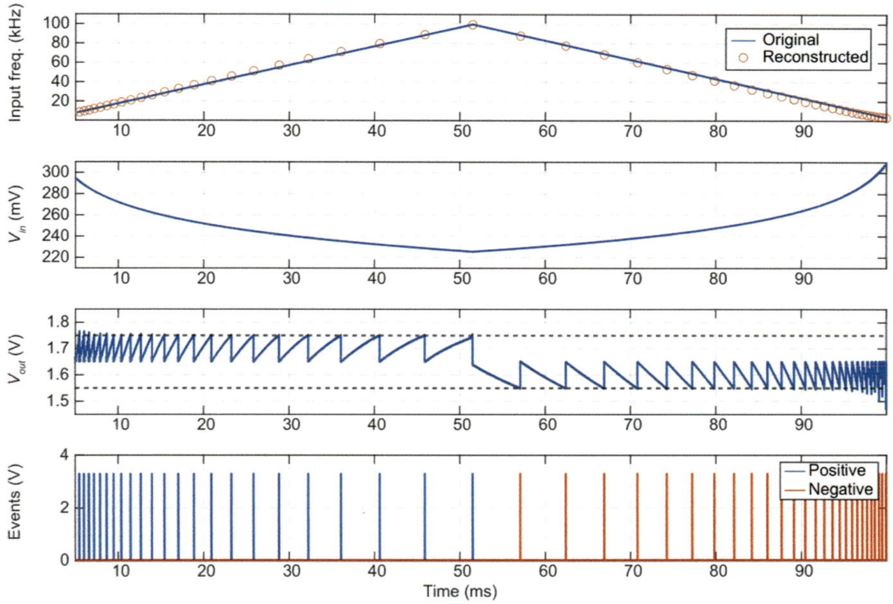

FIGURE 4.22 Simulation results of analog discrete DVS.

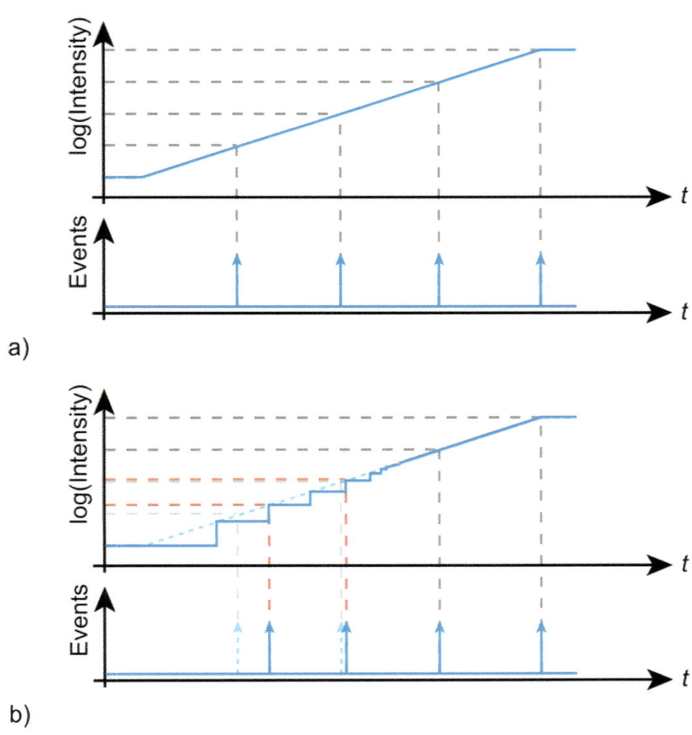

FIGURE 4.23 a) Event generation in a continuous-time DVS. b) Event generation in discrete DVS illustrating errors due to fast transients at low intensity levels.

thresholds at the onset and conclusion of the transient, primarily due to increased latencies when the input frequency is too low.

Although both continuous-time and discrete DVS experience a reduction in bandwidth under dark conditions, the impact is more pronounced in discrete DVS. This difference arises because discrete DVS requires longer times to collect individual photons compared to the continuous accumulation of linear-mode photodiodes. However, this issue can be somewhat mitigated by dynamically adjusting the value of N_{ph} based on the illumination levels.

Regarding the performance of the circuit, all principles outlined in Section 2.3.3 can be applied to this implementation. Despite the simulation using an ideal pulse train whose frequency is not contingent on Poisson statistics, the circuit could encounter false detections and missed variations due to inaccuracies in estimating intensity caused by photon shot noise. Moreover, the operational speed is restricted by the absolute pixel intensity. The TVC is sampled each time the pixel spikes. Consequently, pixels with lower illumination will require more time to accumulate enough photons, leading to increased latency from intensity variation to event detection as the intensity decreases.

However, even in the absence of photon shot noise, this circuit is far from being ideal. Controlling the operation of the circuit within the sub-threshold region is challenging. Moreover, the sample and hold operation poses difficulties. Leakage current in addition to clock feedthrough and charge injection, can result in the loss of memory stored in C_1 and C_2. Additionally, numerous samples in this operation, along with transistors operating in the sub-threshold region, significantly affect noise levels.

5 Characterization of the SAER vision system

This chapter includes experimental results to verify and validate the architectures, the circuits, and the methods presented in previous chapters—behaviors assessed throughout the chapter span from low-level basic detector behaviors to high-level SAER vision system operation modes.

First of all, critical parameters of the SPADs embedded in the SAER sensor are characterized including: spectral response, dark count rate, photon detection efficiency, and jitter characterization. This detailed analysis provides valuable insights into the behavior of SPAD devices under various conditions.

Following this, attention turns towards system-level characterizations. Interactions and dependencies among individual components, including pixels, the sensor, and the overall camera architecture, are meticulously characterized. Controlled experiments facilitate the validation of the sensor across diverse scenarios, providing a quantitative analysis of its performance metrics.

Experimental validation, complemented by simulation outcomes, serves to validate previously discussed theoretical concepts and provides substantial quantitative data. These results act as a significant benchmark, offering insights into the capabilities and limitations of the event-driven SPAD-based 3D image sensor. Additionally, they lay the groundwork for future research, enabling the refinement and advancement of this technology.

5.1 DEVICE CHARACTERIZATION

The performance of the event-driven SPAD-based 3D image sensor fundamentally depends on the quality and characteristics of its photodetectors. In ideal circumstances, a SPAD device would exhibit the ability to detect and respond to the arrival of every single photon without any dead time between detections or the generation of spurious pulses unrelated to photon incidence. Moreover, an optimal SPAD would produce a signal with no temporal variability, commonly referred to as jitter.

In reality, however, SPAD devices deviate from these idealized behaviors, presenting imperfections that necessitate meticulous scrutiny. Several critical parameters assume pivotal roles in evaluating the performance of these devices [Pere18]:

DOI: 10.1201/9781003427490-5

- **Dark Count Rate (DCR):** DCR quantifies the rate at which a SPAD generates spurious pulses in the absence of photon incidence. A lower DCR is indicative of a more reliable device, as it mitigates the likelihood of false positives. **Photon Detection Probability (PDP):** PDP quantifies the probability of a SPAD successfully detecting a photon that impinges upon its active area. Higher PDP values are desirable as they enhance overall sensitivity of the sensor.
- **After-Pulsing Probability (APP):** This parameter addresses the likelihood of a SPAD generating additional pulses after an initial photon detection event. Minimizing after-pulsing is imperative to accurately discern discrete photon arrivals.
- **Jitter**: Jitter is related to the temporal variability in the response time of a SPAD following the arrival of a photon. Minimizing jitter is crucial for precise timing measurements, as it directly affects the accuracy of depth computation.
- **Crosstalk:** SPADs may be influenced by neighboring devices, triggering spurious avalanches related to photon detection of surrounding devices.

The first step in qualifying the SAER vision sensor is characterizing the embedded front-end SPAD detectors.

5.1.1 Breakdown voltage

The first critical parameter in understanding the behavior of a SPAD is its breakdown voltage, denoted as V_{bd}.This parameter dictates the bias voltage that must be applied to the device for proper operation. The SPAD is typically biased slightly above this breakdown voltage by a margin represented as V_{ex}.

Note that V_{ex} plays a substantial role in determining the response characteristics of the device. All parameters described at the beginning of this section are dependent on the chosen value of $V_{ex.}$ Consequently, fine-tuning this parameter is essential for tailoring the performance of the SPAD to specific application requirements.

The breakdown voltage can exhibit variance from one SPAD to another, even within a batch of geometrically identical device instances. Consequently, it becomes crucial to extract statistical data regarding this parameter. This statistical analysis is indispensable when implementing an array of SPADs, as it directly impacts the uniformity of the system, ensuring consistent performance across the array.

To determine V_{bd}, the simplest and most widely employed method involves sweeping the voltage applied to the SPAD device and subsequently extracting the I-V characteristic curve. Then, V_{bd} is identified as the voltage point at which the current experiences a sharp and sudden increase, indicative of the avalanche breakdown phenomenon.

Figure 5.1 (a) shows the I–V curves of a set of 96 SPADs at room temperature, demonstrating this characteristic behavior. Notably, the statistical analysis of the breakdown voltage across this ensemble of SPADs reveals a mean value of 17.96 V, with a standard deviation of 7.7 mV. Figure 5.1 (b) depicts the histogram of the breakdown voltage along this set of SPAD devices. This statistical insight provides essential information for ensuring consistency and reliability in the performance of SPAD arrays.

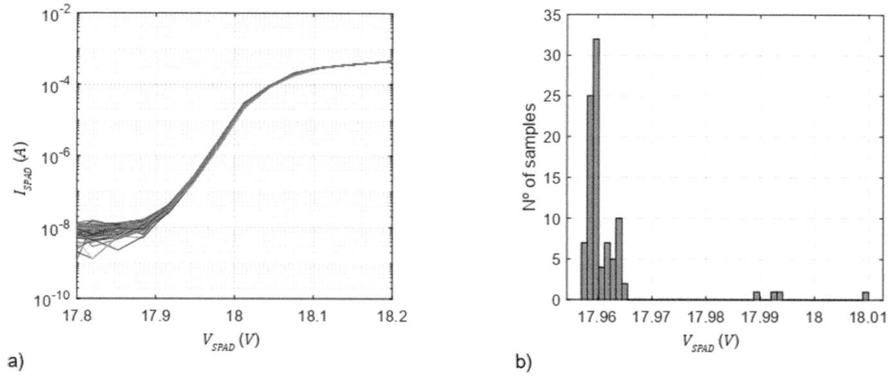

FIGURE 5.1 a) I-V curve of 96 test SPAD devices at = 27°C. b) Breakdown voltage histogram.

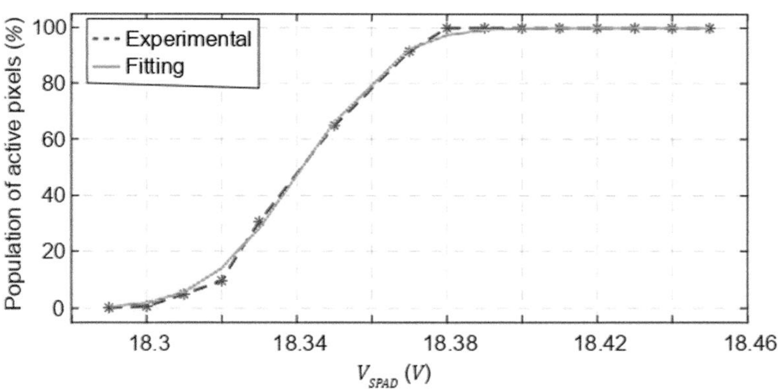

FIGURE 5.2 Cumulative percentage of active pixels vs. applied voltage.

In an additional experiment, the voltage applied to the cathode, V_{SPAD}, was systematically varied, and images were captured at each specific voltage level. This facilitated the determination of the number of pixels operating above the breakdown voltage. The results, graphically depicted in Figure 5.2, represent the cumulative percentage of pixels actively generating data at each voltage increment. Subsequently, the curve was meticulously fitted to the cumulative density function of a Gaussian distribution, yielding a mean voltage of 18.34 V, with an impressively tight standard deviation of 11.1 mV.

This supplementary experiment not only reaffirms the findings illustrated in Figure 5.1, but also provides valuable insights into the threshold voltage of the inverter within the quenching and recharge circuitry. The average threshold voltage is measured at 0.38 V, with a standard deviation of 8 mV. This aligns with the expected values from the initial design phase.

Also, the breakdown voltage of a photodiode is strongly dependent on temperature. As temperature rises, atomic vibrations intensify, consequently decreasing the mean

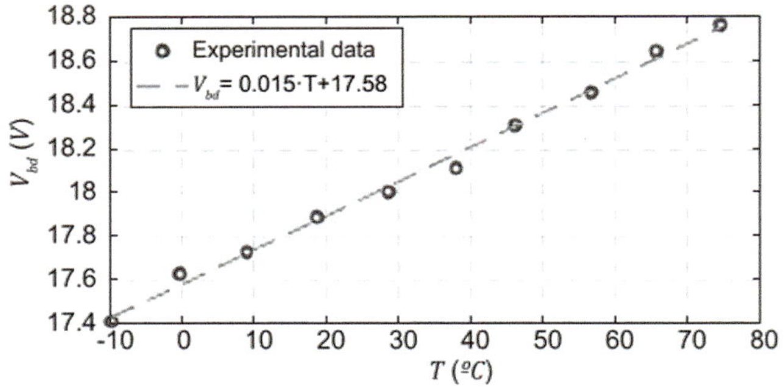

FIGURE 5.3 Breakdown voltage dependence on temperature.

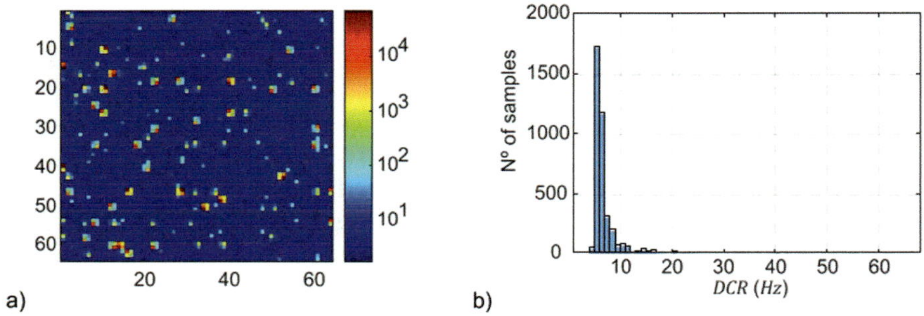

FIGURE 5.4 a) DCR map of a sample of the SAER vision sensor @ Vex = 3 V and T = 27ºC. b) DCR histogram.

free path of electrons [McKe66]. Consequently, the breakdown voltage increases with temperature. Figure 5.3 depicts the dependency of V_{bd} with temperature, exhibiting variations greater than 1 V.

5.1.2 Dark count rate

Measuring the DCR is a straightforward task in SPAD-based sensors that incorporate photon-counting capabilities. Typically expressed in counts per second, obtaining a DCR map requires only capturing an image under dark conditions. This map not only provides information about the absolute DCR, but also yields statistical insights across samples.

Figure 5.4 presents a DCR map at room temperature, illustrating the distribution across the pixel array when $V_{ex} = 3V$. This map facilitates the detection of *hot pixels*, or *screamers*, characterized by an elevated DCR. Notably, these hot pixels generate a higher number of events due to dark counts, a scenario undesirable since it can impact neighboring pixels via crosstalk phenomena and consume a significant portion of the bandwidth of the readout channel. In fact, a 2×2 squared pattern can be detected in

Figure 5.4, indicating that pixels sharing the same cathode with a hot pixel are probably influenced by crosstalk. The median DCR is 5.3 Hz and the standard deviation after removing hot pixels results in 4.5 Hz.

The DCR is strongly dependent on the applied excess bias voltage V_{ex}. Figure 5.5 illustrates the dependence of the median DCR on V_{ex}. This dependence arises from the fact that V_{ex} governs the electric field strength across the active region of the SPAD. A higher V_{ex} leads to an intensified electric field, thereby amplifying the likelihood of spontaneous electron-hole pair generation and consequently, increasing the DCR.

Previous experiments were at a fixed temperature of 27 °C. However, because thermally generated carriers significantly contribute to DCR, characterizing the temperature dependence is essential for practical usage. At higher temperatures, the increased thermal energy facilitates more frequent thermally induced carrier generation, leading to a higher DCR. Trap-assisted and thermal generation are well-described by the Shockley-Read-Hall recombination model [Shoc52]. This model explains how carriers recombine in the presence of defects within the semiconductor material. These defects act as recombination centers, providing an alternative path for carriers to recombine, which in turn contributes to the DCR.

Conversely, lower temperatures reduce the rate of thermally generated carriers. In fact, as temperature decreases *Band-to-Band Tunneling* (BTBT) [Kane61] dominates over thermally generated carriers. This mechanism enables carriers to traverse the energy gap between the valence and conduction bands, which contributes to the DCR.

The generation processes exhibit an exponential dependency on 1/T, making the Arrhenius plot a more suitable representation. Consequently, the trend for carrier generation and, consequently, the dependency of DCR on temperature is as follows [Piem19]:

$$DCR = Ae^{-\frac{E_A}{kT}} \tag{5.1}$$

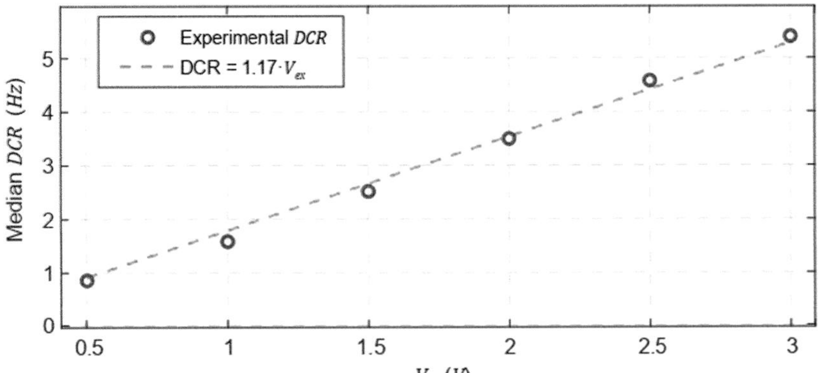

FIGURE 5.5 Dark count rate as a function of the excess voltage @ T = 27 °C.

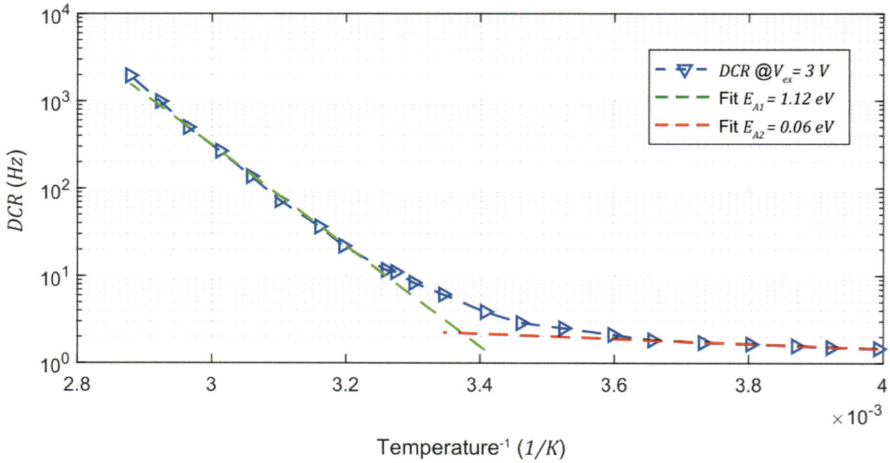

FIGURE 5.6 DCR as a function of temperature @ Vex = 3 V. Ea1 is the slope associated to thermal trap assisted, while Ea2 is the slope associated to band-to-band tunneling.

where A is a constant and E_a is the activation energy of the charge carriers that trigger the spurious avalanches. Figure 5.6 illustrates the temperature dependence of the SPADs within the SAER vision sensor, highlighting the regions where thermally generated and band-to-band tunneling carriers dominate. Note that the absolute value of V_{bd} must be adapted to ensure a constant excess voltage.

Figure 5.6 suggests that the diffusion component dominates at high temperature, since the activation energy fits with that of silicon. On the other hand, at low temperature, BTBT dominates and the DCR starts to plateau. Small dependencies in the latter regime are attributed to temperature variations of the silicon bandgap with temperature [Sze06].

5.1.3 Photon detection probability

PDP corresponds to the likelihood of successfully detecting an incident photon with a specific energy, typically associated with a particular wavelength. The methodology employed for PDP estimation is similar to that of conventional photodiodes. In our investigation, we utilized a dedicated experimental setup, using a calibrated light source and a monochromator.

A calibrated light source is essential to accurately determine PDP. The spectral information from the SPADs was extracted using the monochromator integrated in series with the light source, allowing us to filter the light with a specified bandwidth of 10 nm. The output of the monochromator is then directed into an integration sphere. By measuring the resulting optical power output per area unit, P, we can subsequently calculate the average photon arrival rate, ϕ, as follows:

$$\phi = \frac{P\lambda}{hc} A \tag{5.2}$$

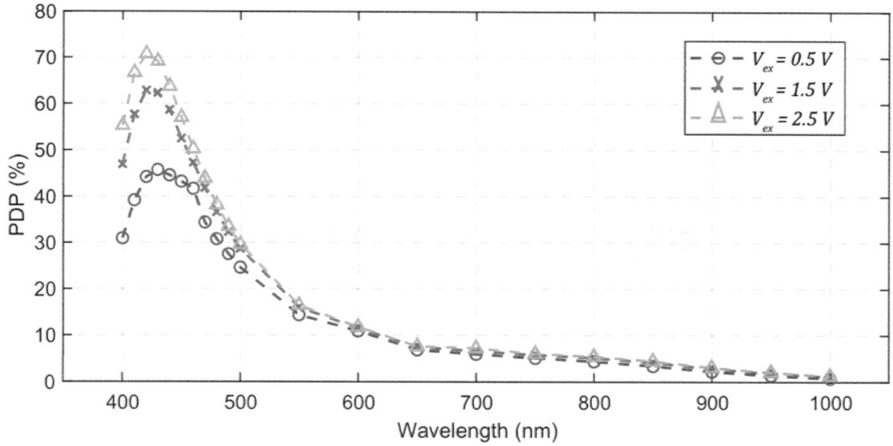

FIGURE 5.7 Photon detection probability as a function of the wavelength.

where A is the active area of the detector, λ is the wavelength of the incoming photons, h is the Planck's constant, and c is the speed of light.

Next, we can calculate the PDP by quantifying the average number of detected photons. It is crucial to note that the optical power must be carefully selected to operate within the single photon regime, typically around 1 nW/mm^2 to avoid pile-up effects that may otherwise alter the measurements.

Figure 5.7 illustrates the PDP of SPAD devices concerning various wavelengths. Notably, the peak occurs at 420 nm, achieving a PDP exceeding 70% under sufficiently elevated excess voltage. This value is relative large when compared to the state-of-the-art [Gram22; Leit13; Nicl07; Veer15; Veer15]. The peak PDP with respect to previous implementations [Vorn21] may be attributed to the blocking of the silicide process in the active region of the junction [Chen16]. Conversely, the PDP significantly diminishes to well below 5% in the NIR range, as indicated in Figure 5.8. Since the DCR increases at a higher rate with V_{ex} than the PDP, a low value of V_{ex} is of interest, e.g., 0.5 V. However, a trade-off between sensitivity and dynamic range appears due to this increase in the noise floor marked by the DCR.

5.1.4 After-pulsing probability

After-pulsing is an inherent effect in SPADs where, subsequent to the initial detection of a photon, an additional pulse is generated that is not correlated with an actual photon arrival. This phenomenon arises from the trapping and subsequent release of carriers during the avalanche process, leading to a secondary triggering event [Pere18].

This effect is highly undesirable as it introduces spurious signals that can distort accurate measurements. To mitigate after-pulsing, the dead time of the SPAD can be configured to be sufficiently long, effectively avoiding trap-assisted avalanches and minimizing the occurrence of after-pulses [More19].

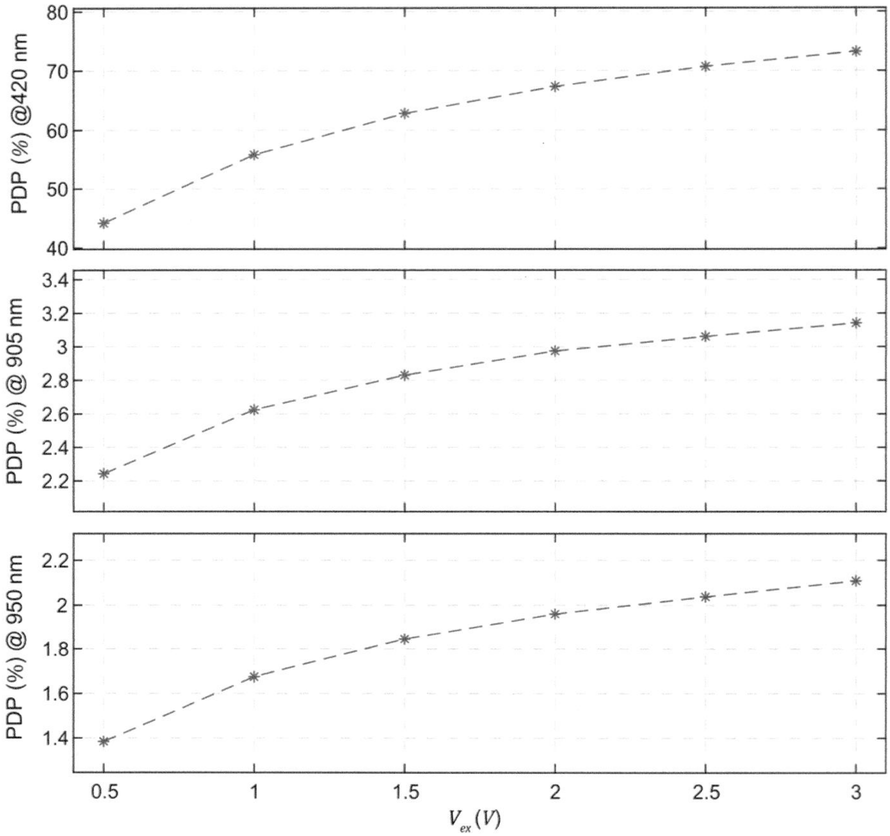

FIGURE 5.8 Photon detection probability as a function of excess voltage.

A common approach to quantifying APP involves measuring the inter-arrival time between avalanches. In a Poisson process, the distribution of inter-arrival times follows an exponential distribution. Therefore, under dark conditions, the behavior of a SPAD is conditioned by DCR. By comparing the experimental inter-arrival time distribution to the fitting of an exponential curve, the APP can be extracted.

Figure 5.9 present the histograms of inter-arrival times for a hot pixel with DCR = 169.8 kHz biased at $V_{ex} = 3V$. The results suggest that the effect of after pulsing is negligible since no deviation from the exponential fitting is observed. This result suggests that the dead time can be configured to be as low as possible.

Nevertheless, this measurement scenario is somewhat idealized. The experiment illustrated in Figure 5.9 involved the activation of a single pixel during the measurement. However, when all pixels are enabled simultaneously and the dead time is reduced, avalanches occurring in closer succession might appear across the array. Consequently, higher current peaks might potentially lead to noticeable voltage drops in the power supply. Such voltage fluctuations can significantly impact the operation of the transistors within the quenching and recharge circuit. In extreme cases, it

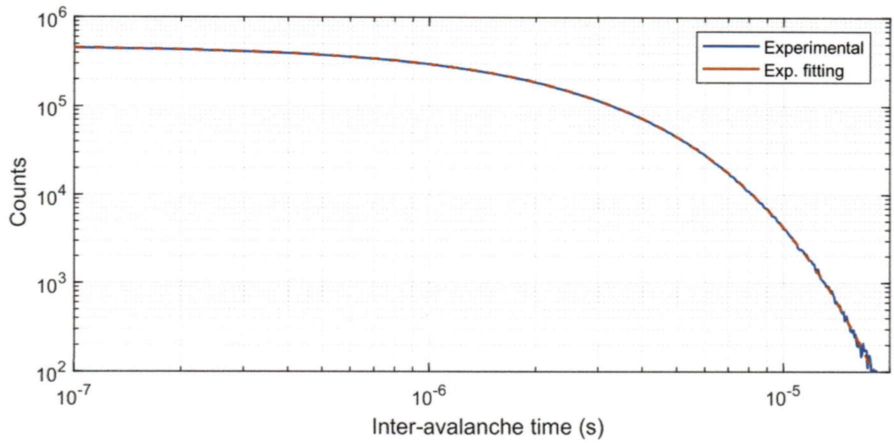

FIGURE 5.9 Histogram of the inter-arrival time between consecutive pulses.

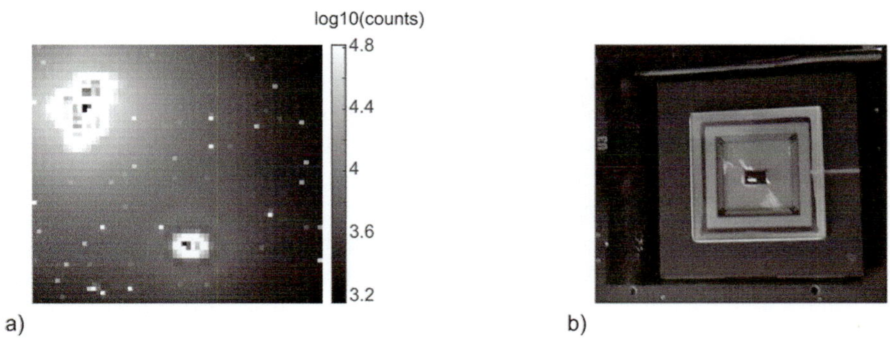

FIGURE 5.10 a) Image under dark conditions showing illuminated spots due to a short dead time. b) Picture of the sensor emitting light generated by avalanche photoemission.

may even result in a scenario where the reset transistor is not completely turned on, providing a path for the flow of avalanche current. Another possible reason is that decreasing the dead time involves increasing the current of the in-pixel delay, thus, a voltage drop at the reference node due to a higher current can also impact the strength of the reset transistor.

Figure 5.10 (a) shows an image captured during dark conditions. The image reveals two prominent spots, indicating that the central black pixels are those in avalanche conduction, emitting photons subsequently absorbed by neighboring pixels. Remarkably, these pixels exhibit a non-zero output, suggesting they have not been emitting photons since the beginning of the operation. Instead, there is an instant when the SPAD enters conduction, leading to photon emission, while the neighboring pixels remain active for the entire operation. This aligns with the hypothesis that a voltage drop induces the SPAD into a conduction state, a concept explored further in

Section 5.2.1. Figure 5.10 (a) illustrates a picture of the sensor emitting light due to avalanche photoemission.

Therefore, the minimum dead time of the sensor is selected to be 1 μs. Under this condition, the sensor does not show any undesired effect, at the expense of a lower dynamic range.

5.1.5 CROSSTALK

Crosstalk in SPAD-based sensors refers to the interference and disturbances originating from neighboring elements within the device. These interactions can have a significant impact on the accurate detection of photons.

There are various sources of crosstalk in SPAD-based sensors. Firstly, there is electrical crosstalk arising from capacitive coupling between adjacent devices. Additionally, every 2×2 SPADs share a common terminal (cathode), which can lead to electrical crosstalk.

Optical crosstalk is another form, stemming from the emission of secondary photons during the avalanche phase. These secondary photons may trigger neighboring SPADs, introducing unwanted signals.

To quantitatively assess crosstalk within this sensor, individual SPADs can be selectively disabled, allowing for a comparison between the average count rates with the device enabled and disabled. Figure 5.11 provides a visual representation of the crosstalk probability between neighboring SPADs, demonstrating negligible crosstalk with SPADs that do not share a common cathode. Besides, the higher crosstalk probability (1.93%) corresponds to the SPAD sharing a longer edge.

Understanding and mitigating crosstalk is crucial for achieving accurate and reliable photon detection in SPAD-based systems, particularly in scenarios where high precision and low noise are essential. This consideration ensures that measurements are not excessively influenced by interference from neighboring elements.

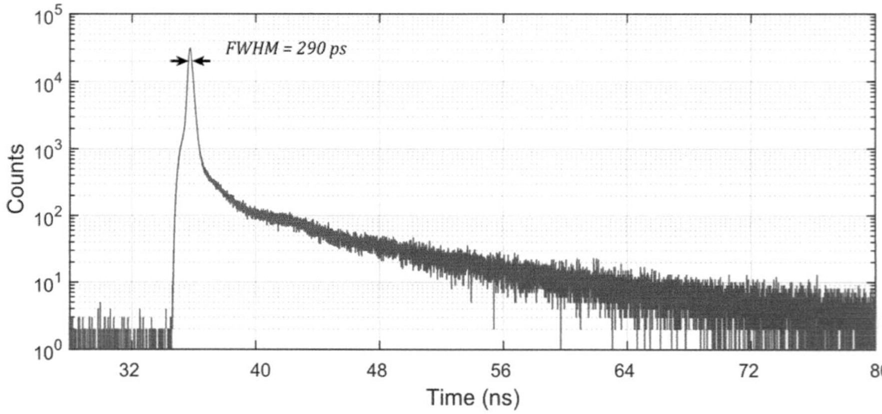

FIGURE 5.11 Crosstalk map of SPAD among neighboring devices. The SPAD highlighted in red correspond to the aggressor.

5.2 PIXEL CHARACTERIZATION

This section presents an analysis of the pixel, which stands as the fundamental unit responsible for initial data processing, specifically handling incoming photons. Within the pixel, two primary functions are executed: the integration of photon counts, and the retention of photon counts from prior exposures. Depending on the operational mode of the pixel, events are generated and subsequently transmitted outside the pixel array. As such, factors such as speed and power consumption are crucial, presenting a profound influence on the scalability and viability of the proposed sensor technology.

This study delves into the intricate temporal dynamics governing the operation of the pixel. Multiple time delays play a significant role, including mechanisms involving quenching and recharging, counter bit flips, memory writing, and event transmission. Understanding and quantifying these temporal intricacies is essential for fine-tuning the performance of the sensor, ultimately ensuring its optimal functionality within diverse operational contexts.

5.2.1 QUENCHING AND RECHARGE CIRCUITRY

As explicated in Section 1.1.1, the operation of the SPAD initiates from a stable state. However, following photon detection, the avalanche must be promptly quenched, and subsequently, the SPAD voltage requires recharging. It is crucial to note that these operations are not instantaneous. Consequently, the performance of the SPAD is constrained by the duration required to extinguish an avalanche and the time for voltage restoration.

The quenching and recharge circuitry additionally assumes the responsibility of imposing a dead time between photon detection and voltage restoration. Ideally, the quenching time should be minimized to mitigate the duration in which the SPAD draws a large current, while the recharge time should also be minimized to reduce the time between detection of subsequent photons.

However, the first aspect when designing the quenching circuitry is that its resistance is large enough to extinguish the avalanche. Otherwise, a SPAD can continuously drive a large current, increasing its temperature and decreasing its breakdown voltage, contributing to a positive feedback loop where the current can increase. Figure 5.12 depicts the testbench we implemented to verify the quenching and recharge circuitry. The bias voltage V_{bq} is generated from a programmable current source, which bias M_{Q1} to prevent it from driving a current larger than I_{bq} when it enters saturation region. Then, the current flowing into the circuit, I_{SPAD}, was swept to find the current that triggers the quenching circuit, i.e., the minimum avalanche current that the circuit can extinguish.

Figure 5.13 shows the relationship between I_{SPAD} and the anode voltage, V_a, for different I_{bq} values. The results indicate that the point at which V_a experiences a sudden increase, i.e., the current threshold, I_{th}, at which an avalanche is detected and extinguished, corresponds to I_{bq}. However, with increasing I_{bq}, this point tends to plateau. This is primarily due to M_{Q1} transitioning to behave like an ideal switch, with the resistance being primarily determined by M_{Q2}. Note that above this current threshold the voltage settles at the operating point where the current flows through

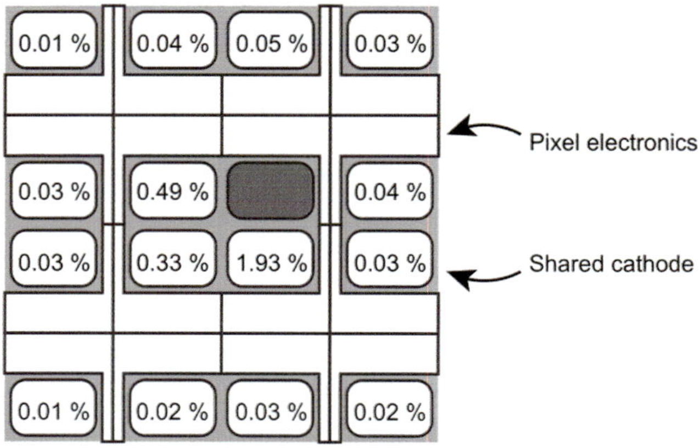

FIGURE 5.12 Testbench employed for characterizing the quenching circuitry.

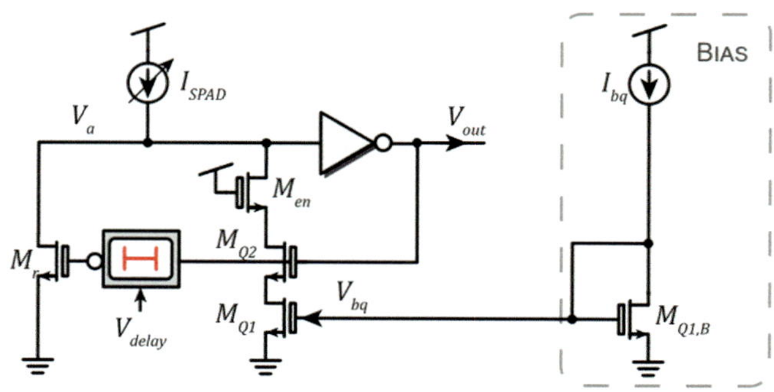

FIGURE 5.13 Anode DC voltage as a function of ISPAD. Points where the voltage suddenly increases correspond to the current threshold where an avalanche is extinguished.

the parasitic junction of M_{en}. However, this scenario does not manifest in practice, as the current diminishes with the increase in V_a. Figure 5.14 depicts I_{th} as a function of I_{bq}.

Another crucial aspect is related to the rise time. During the design phase, a simulation was conducted using the SPAD model from [Lope18], with model parameters scaled from a prior design within the research group [Vorn21]. Figure 5.15 illustrates the rise time as a function of I_{bq}. As the conductance of M_{Q1} increases, the equivalent resistance decreases, resulting in a longer rise time, which is typically undesirable. Although the absolute value of the rise time depends on the actual parasitic capacitances, an estimate of its relative variations can be extracted from the simulation, indicating a 55% increase in rise time. Consequently, the bias point of M_{Q1} must be carefully selected to enable the circuit to terminate the avalanche with the shortest rise time. This is achieved with the lowest value of I_{bq}; however, it also increases the noise contribution due to the increased resistance and, thus, jitter. In terms of

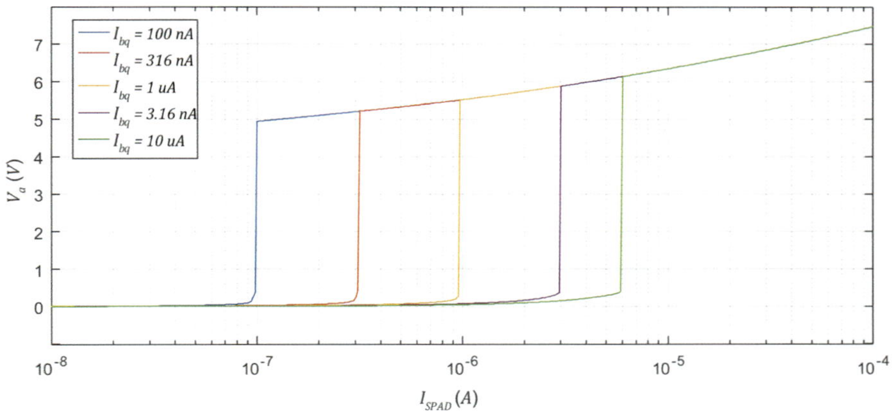

FIGURE 5.14 Evolution of the current threshold as a function of I_{bq}.

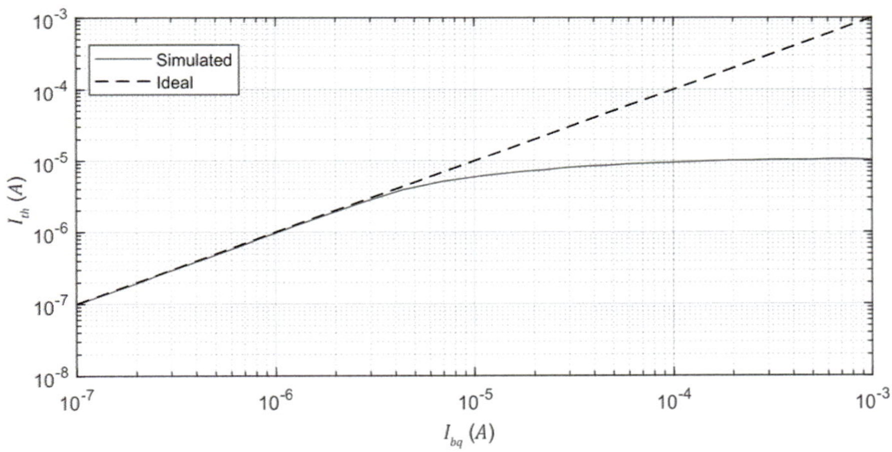

FIGURE 5.15 Simulated rise time of the anode voltage as a function of I_{bq}.

variability, Monte Carlo analysis reveals that I_{th} exhibits a standard deviation less than 10% of its average value at low I_{bq} values. These variations drop below 1% for I_{bq} values exceeding 10 μA. Consequently, statistical variations are not particularly significant, as the rise time barely fluctuates at low currents, and negligible variations in I_{th} emerge at higher currents.

The dead time of the SPAD was measured in the laboratory to determine the programmable range. Figure 5.16 shows the evolution of the dead time versus V_{delay}, showing that the minimum dead time that can be achieved is 10 ns and the maximum is greater than 20 μs. Monte Carlo simulation shows variations below 5% when aiming for a delay of 1 μs, which are reduced as the delay increases.

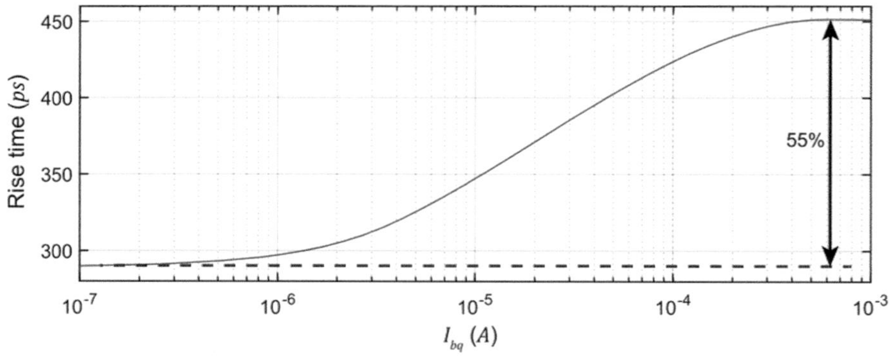

FIGURE 5.16 Experimental dead time of SPADs as a function of V_{delay}.

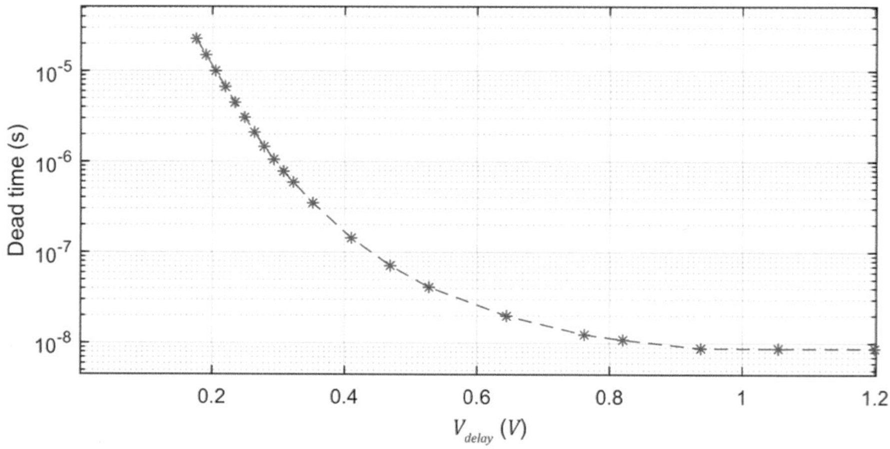

FIGURE 5.17 a) Critical nodes affected by a power supply drop. b) Corresponding waveforms.

However, the implemented quenching and recharge circuit also exhibits certain limitations, as previously discussed in Section 5.1.4. While passive quenching ensures proper operation provided its resistance is sufficiently high to effectively quench the avalanche, potential issues may arise in this implementation.

When the sensor encounters a high activity within a specific time, any decrease in voltage at the power supply or reference node results in reduced voltage levels of control signals. Figure 5.17 (a) showcases the critical nodes affected by a voltage drop occurring at the 1.2 V power supply.

Figure 5.17 (b) depicts a waveform diagram illustrating a problematic scenario that can compromise the operation. In the event of a voltage drop, V_r decreases, hence, the strength of M_r also does, leading to a longer recharge time. Consequently, the likelihood of detecting a new photon during the recharge phase increases. In such cases, there exists a probability that M_r reaches a stable point where its current matches the avalanche current. This situation could potentially induce metastability in the inverter, which can be propagated and can keep the SPAD in conduction even when the power supply voltage is restored. However, metastability elevates the current consumption of the pixel and contributes to the voltage drop. This positive feedback loop intensifies as the photodiode begins emitting photons, detected by neighboring pixels, thereby increasing their activity and power consumption.

While this behavior may not be easily observable from outside the sensor, several reasons support this hypothesis. Firstly, when the sensor is disabled (*enable* off), the output inverter and the recharge branch remain inactive, operating the circuit in a passive quenching mode. In this scenario, no SPAD enters conduction, thereby avoiding high current consumption and photoemission. Secondly, SPADs enter conduction when an extremely illuminated spot appears in the scene during operation. Additionally, reducing the dead time leads to increased current peaks in the pixel and the potential for consecutive avalanches. Furthermore, decreasing the excess voltage diminishes the effect, aligning with the hypothesis, as it results in a reduced excursion in V_a and reduced data rate.

In conclusion, the active quenching and recharge circuit implementation offer several advantages. The rapid quenching mechanism, facilitated by positive feedback in the quenching branch, restricts maximum current, decreasing SPAD power consumption. This reduction in current and time lowers the probability of carriers being trapped in intermediate states, mitigating after-pulsing risks. Moreover, the quick recharge phase ensures stable pulse width and dead time, enabling precise control over pile-up effects. However, this design occupies considerably more area compared to a passive quenching and recharge circuit, which could be implemented using a single transistor or poly resistor. Additionally, as detailed in this section, separate quenching and recharge branches may reach stable states where quenching remains incomplete.

5.2.2 Jitter

Another crucial parameter in SPAD devices is jitter. It refers to random variations relative to a reference time point, commonly linked to deviations in the timing of avalanche occurrences with respect to photon arrivals. Maintaining low jitter at the device level is imperative, as it directly affects the perceived pulse width, potentially worsening temporal resolution.

Jitter can be influenced by the physical size of the SPAD device, as a larger area may result in varying resistances at different points of the SPAD to the contact, and more different injection positions. Indeed, an avalanche initiated at the center of the active area experiences faster growth compared to one initiated near the border. This disparity arises due to the varying lengths of the effective wavefronts in these regions [Spin97]. Alternative studies have also demonstrated that injection point is not necessarily the main contributor to jitter [Assa11]. Additionally, the quenching circuitry plays a role in jitter, as electrical noise can disrupt the avalanche process.

For accurate measurement, jitter is typically assessed through time-correlated single-photon counting [Bron16]. By employing a well-characterized light source,

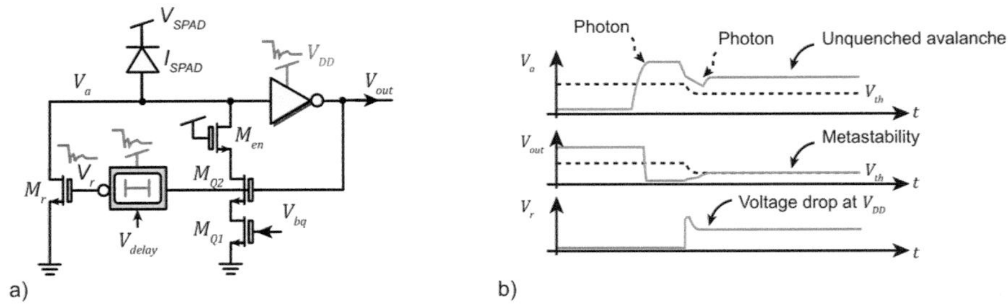

FIGURE 5.18 Time-interval histogram collected with 4 ps resolution. Full width at half maximum is 290 ps.

such as a picosecond pulse laser, a histogram is constructed. Jitter manifests as an increase in pulse width, allowing its determination by subtracting the *Full Width at Half Maximum* (FWHM) of the laser pulse from the FWHM of the histogram.

Figure 5.18 illustrates a resulting histogram generated using a PicoQuant laser. However, as the manufacturer provides a range (70 ps to 300 ps) rather than an exact FWHM, corrections cannot be made. Consequently, the jitter of both the SPAD and quenching circuitry is confirmed to be less than 290 ps.

5.2.3 Counter delay

Following the quenching and recharge circuitry, the counter constitutes the subsequent unit in the processing chain. However, it introduces a delay between the initial detection of a photon and the eventual transmission of an event. This temporal gap arises due to the inherent structure of the counter, wherein the number of stages directly impacts the duration of the delay. Therefore, since signals are encoded in the time domain, any delay introduced during the generation and transmission of events must be considered.

It is important to note that measuring parameters inside the pixel is a notably challenging task. The circuits within the pixel are designed for high-speed operation, resulting in delays in the tens-to-hundreds of picosecond range. Accurately measuring these signals presents several difficulties. Firstly, inserting buffers to drive parasitic loads related to pads, interconnections, and package and PCB parasitics can significantly load the circuit. Additionally, these parasitics impose limits on the frequency of the output signal, posing a challenge for single-ended signals. Moreover, when measuring relative times in the picosecond range, any asymmetry in the load of two signals can considerably corrupt the results. As a result, some metrics are restricted to either indirect measurements or simulation results.

To obtain the most faithful results from simulations, it is imperative to include extracted parasitics. Figure 5.19 illustrates the delay from the output of the quenching circuitry to each output bit of the counter, being the average delay introduced by each flip-flop, $t_{clk \rightarrow Q}$ is 332 ps. It is important to consider that, as the output bit b increases, the number of photons per event grows exponentially as 2^b. Consequently, the effect of delay on the measurement is diminished. Additionally, the only delay influencing

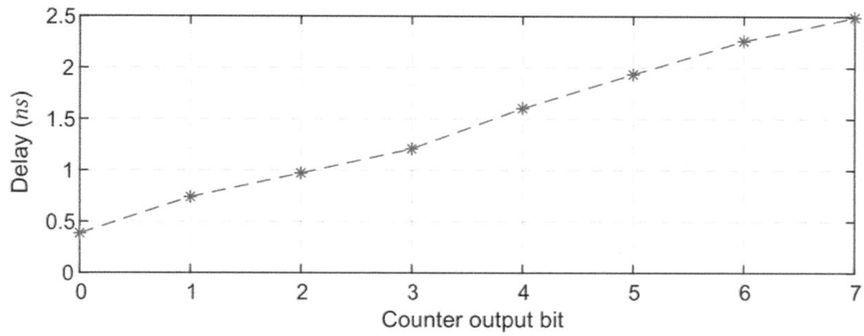

FIGURE 5.19 Simulated delay from counter input to each output bit.

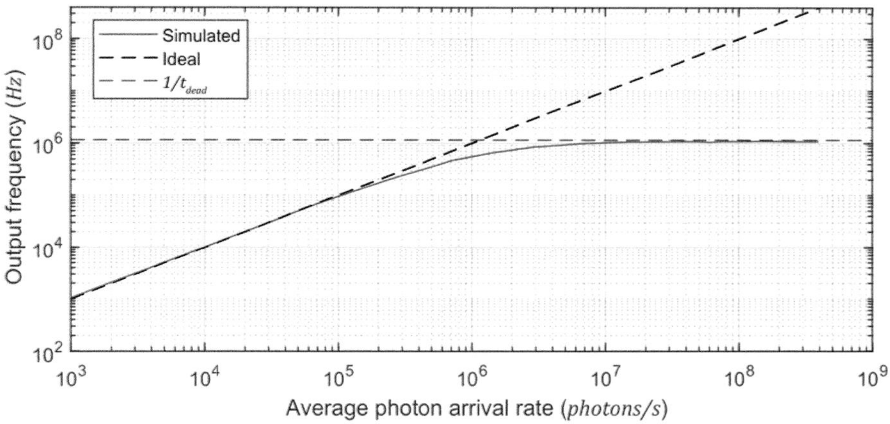

FIGURE 5.20 Output frequency of the first bit of the counter as a function of the average photon arrival rate. Maximum frequency is limited by the dead time of SPADs.

functionality is that of the first stage, i.e., the first flip-flop, as the frequency of subsequent stages is reduced. The period of the input signal must be long enough for the data to transition and remain stable during the setup time, t_{setup}, which was measured to be 217 ps in the extracted simulation. This determines the maximum inter-arrival time, T_{min}, which can be expressed as:

$$T_{min} = t_{clk \rightarrow Q} + t_{setup} \tag{5.3}$$

In this pixel, T_{min} is 449 ps. This is not a problem in the current implementation, since $t_{dead} \gg T_{min}$. However, if the pixel includes more than one SPAD to address this limitation, this time must be considered when determining the width of the pulses corresponding to each photon detection.

Figure 5.20 shows the simulated output frequency of the counter when its output is selected to be the first bit (b=0). In this simulation, the Poissonian behavior of photons was considered, assuming that 100% of the photons are detected. As can be observed, the maximum event generation frequency saturates at $1/t_{dead}$.

5.2.4 Multiplexer delay

The multiplexer within the pixel also contributes to delay. In the SAER pixel, the multiplexer is implemented with analog switches to minimize area consumption. However, this approach means that the signal is not regenerated at each stage of the multiplexer. Thus, the multiplexer is equivalent to an RC circuit, where R is limited by the on-resistance of the switch and C represents the parasitic capacitance at the output of each stage of the multiplexer. This includes both the junction capacitance of the switch and the interconnection parasitics. Consequently, the input signal sees a low-pass filter of which the output delay, $t_{d,mux}$, can be estimated using the Elmore delay [Elmo48]:

$$t_{d,mux} \simeq RC\frac{N(N+1)}{2} \tag{5.4}$$

where N is the number of stages in the output multiplexer. Extracted simulations showed that the total delay introduced by the multiplexer is 594 ps. Thus, the error introduced by the counter and the multiplexer when the sensor operates in octopus or TNP mode can be expressed as:

$$\epsilon_{max} = \frac{(b+1)\cdot t_{clk\rightarrow Q} + t_{d,mux}}{2^b \cdot T_{min}} \tag{5.5}$$

yielding an error of 5.65% at $b = 7$ when T_{min} is limited by the counter and 0.0025% when T_{min} is limited by the dead time (1 μs).

5.2.5 Pixel operating in the linear mode

To enhance the dynamic range of the SAER pixel and overcome limitations related to dead-time, photodiodes can be biased in the linear regions. While the sensor lacks the capability to selectively bias pixels in distinct areas, a multiplexed operation of the sensor facilitates information extraction from brighter regions.

Section 4.2.1 described that when the photodiode operates below the breakdown voltage and the quenching branch remains inactive, holes are accumulated in the anode of the photodiode, increasing its voltage linearly in response to the illumination. Eventually, the inverter, functioning as a comparator, switches its output, triggering the counter and the recharge branch, which resets the accumulated charge in the photodiode. The behavior depicted in Figure 5.21 illustrates this phenomenon for two levels of illumination, setting N_{ph} to 4 prevents the readout channel from saturation.

An observation from this example is the consistent reset and threshold voltages for the two specific illumination levels. This consistency indicates that the reset transistor is strong enough to reset the voltage regardless of the illumination on the one hand, and that the inverter is fast enough to toggle even for fast input transients on the other,

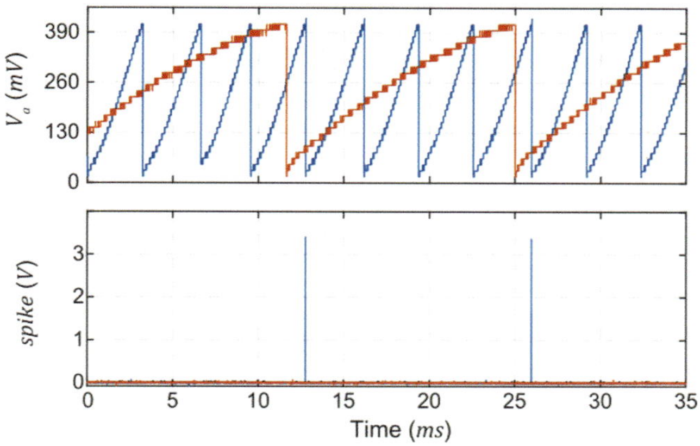

FIGURE 5.21 Waveforms of the SAER pixel operating in the linear mode (I&F).

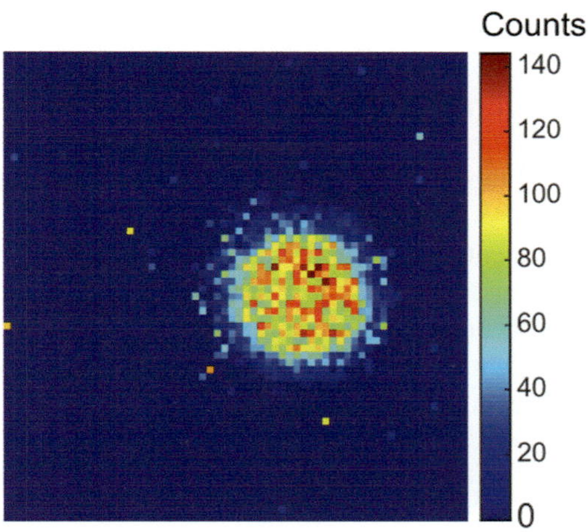

FIGURE 5.22 Rendered image in linear-mode.

avoiding measurement distortion. This particular pixel exhibits an approximate 390 mV excursion in the photodiode. This low excursion is due to the fixed threshold voltage of the inverter, presenting a significant challenge in terms of noise and FPN [Gome23a].

In Section 5.1.1, the standard deviation of the threshold voltage was measured at 8 mV, with an average value of 380 mV. Given that the signal in the voltage domain corresponds to the excursion range, and hence, to the threshold voltage of the inverter, the resulting FPN across the array due to this mismatch is calculated to be 2.1%.

However, due to the reset transistor being designed with high strength for the rapid recharge of SPADs, mismatches in its subthreshold current contribute to an increased FPN to a level that renders it unsuitable for operation within this range. Notably, this subthreshold current imposition restricts the detection of the minimum measurable photocurrent. This limitation is evident in Figure 5.22, which displays an image captured under typical office lighting conditions, specifically featuring a light spot from a mobile phone flashlight. Within this scene, pixels situated outside the bright spots exhibit no output signal. Therefore, optimization of the quenching and recharge circuitry becomes imperative if the sensor is intended to operate effectively in this mode.

5.3 SENSOR CHARACTERIZATION

This characterization is critical in accurately defining the sensor's overall performance by analyzing its various constitutive subblocks beyond the pixel array. Specifically, we will study the AER readout circuitry, the programmable delay, and the biasing circuitry.

Gaining insights into the circuit performance of these components is pivotal. Understanding the constraints related to speed in the readout and power consumption allows us to extract key parameters such as the maximum achievable data rate and the energy expended for each conversion. This information serves as a fundamental benchmark for assessing the efficiency and effectiveness of the sensor in real-world operational scenarios.

Additionally, the precise performance of the programmable delay plays a crucial role in determining the efficacy of the ToF mode. Any non-uniformity or time uncertainty in this component directly contributes to measurement uncertainty. Therefore, a thorough evaluation of the delay circuitry is essential for accurate and reliable ToF measurements.

Finally, the establishment of an appropriate biasing scheme is imperative to ensure consistent and reliable sensor performance. To this end, we conducted measurements to assess the variation of on-chip references under PVT variations. This investigation provides critical insights into the robustness and stability of the biasing circuitry under varying operational conditions.

5.3.1 AER READOUT CIRCUITRY

In Section 5.2, we characterized the delay associated with event generation. However, this parameter alone lacks meaningful context without an understanding of the time required for a pixel to transmit the event and release the readout channel. In essence, we must ascertain the duration it takes for the request signal to transition from a logic high to low, and subsequently return to logic high.

A method to characterize the bandwidth of the readout channel consists of short-circuiting the request signal (bus_req) and the acknowledge signal (bus_ack). In this configuration, each request is answered with the minimum delay imposed by the PCB and package parasitics.

Figure 5.23 provides a graphical representation of bus_req when an event is generated. Notably, the implementation of adaptive pull-up elements described in Section 3.4.2 reduces the required time for event readout. Consequently, we observed a reduction of 29.7% in delay when compared to the scenario where pull-up elements are implemented with PMOS current sources (or resistors), signifying a substantial enhancement in operational efficiency.

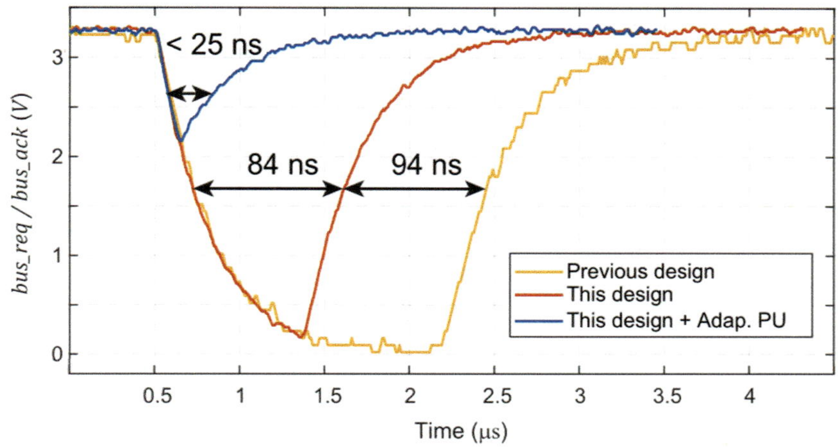

FIGURE 5.23 Event readout time extracted from bus_req signal when connected to bus_ack.

TABLE 5.1
Summary of delays within the AER readout block.

Delay	Description	Value
$t_{r,i-a,i}$	Time from $req\_row[i]$ to $ack\_row[i]$ assertion.	5.1 ns
$t_{a,i+1}$	Time from $req\_row[i]$ deassertion to $ack\_row[i+1]$ assertion given that $req\_row[i+1]$ was already activated. This value represents the shortest delay.	364 ps
$t_{a,i+N}$	Time from $req\_row[i]$ deassertion to $ack\_row[i+N]$ assertion given that $req\_row[i+N]$ was already activated. This value represents the longest delay.	4.5 ns

However, pull-up elements are not the sole contributor to this delay. The arbitration circuit introduces a delay during the arbitration process. Notably, this delay is non-deterministic, contingent upon the prior state of the arbitration process—specifically, which row or column has been granted access to the readout channel. This variability arises from the fact that a portion of the path may have already been granted, requiring fewer arbitration stages for acknowledgment. While this arbitration delay exhibits signal dependence on the scale of the pixel array, it is pertinent to recognize that it also introduces a stochastic element, influencing the noise characteristics inferred from the image.

Simulation results at the worst-case corner are summarized in Table 5.3. They show that the time it takes for a request to be acknowledged is 5.1 ns. However, the delay in attending the next pixel when collision exists is 320 ps for the adjoining element and 4.5 ns for the farthest one. The time difference arises because the lower-level arbiter is granted access by the upper levels when the first element is read out. In Figure 5.24, all elements request access simultaneously. After r_0 releases its request,

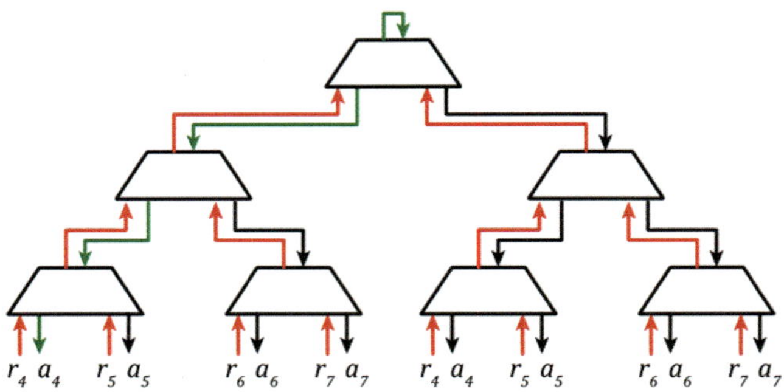

FIGURE 5.24 Sample arbiter tree, representing the granted path and requested path in green and red colors, respectively.

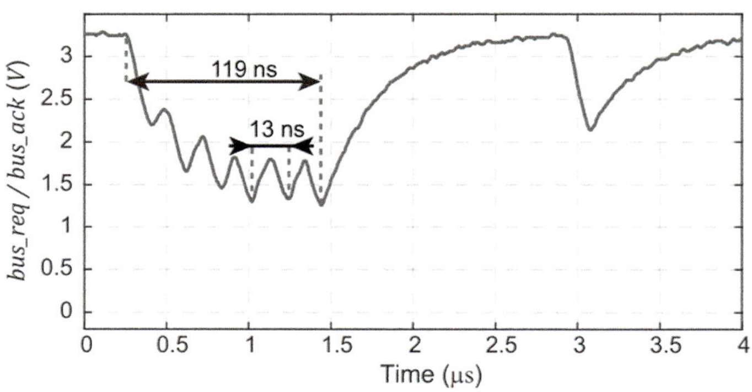

FIGURE 5.25 Event readout time extracted for events in the same row.

r_1 is automatically granted access since the remaining path has already been granted, eliminating the need to wait for the upper levels again. Consequently, the time to the next access is minimized. Conversely, r_{4-7} must wait for the signal to propagate through the entire tree, resulting in the maximum delay. Regarding the SRAM writing operation, the worst-case corner shows that the writing time is 670 ps.

Figure 5.25 showcases the waveform of the signal *bus _ req* when multiple events are readout. Events in the same row do not have to repeat the handshaking protocol at the row level, benefiting from a shorter readout time. In this case, the shortest time between events is 13 ns, while the time required to read six consecutive events is 119 ns, resulting in an average time of 19.8 ns. This representation clearly demonstrates the maximum achievable event rate is approximately 5×10^7 *events per second* (eps).

5.3.2 Programmable Delay

As explained in Chapter 3, the enable gate for pixels serves as a crucial component in the definition of temporal bins from which ToF measurements are subsequently

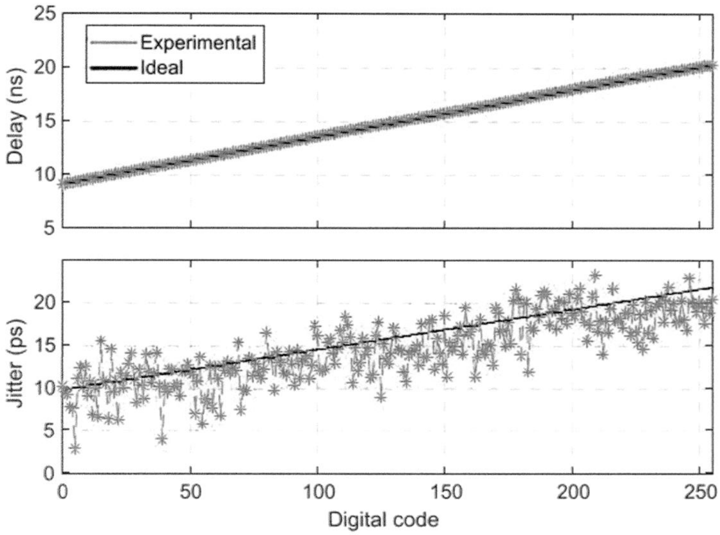

FIGURE 5.26 Output of the programmable delay as a function of the digital code.

derived. Accurate characterization of this circuit is paramount in understanding the limitations that govern the computation of ToF.

We employed a 40 GSa/s oscilloscope with a bandwidth of 13 GHz to measure the delay from the signal that triggers the laser to the signal that enables the sensor. By incrementally increasing the output capacitance of the programmable delay, we measured both the absolute delay and jitter. The jitter of the test pad was measured and subtracted from the latter measurement. Figure 5.26 depicts the delay and jitter for each bin, for the case of the maximum allowable current, i.e., the finest resolution. The maximum jitter reaches around 23 ps, while the delay can be programmed across 256 different positions, each offering a resolution of 44 ps.

Figure 5.27 shows the differential non-linearity and the integral non-linearity of the programmable delay. The experimental results demonstrate that no digital code is lost, and the programmable delay exhibits a monotonic behavior. Also, jitter follows the expected tendency described in Section 3.5, where the dominant contributor is the noise from the power supply. Parameters extracted from the electrical model were employed to evaluate the theoretical model.

Another critical aspect of the programmable delay is the width of its output pulse. Figure 5.28 depicts the pulse width of the *enable* signal as a function of the digital code, ranging from 3.75 ns to 18.35 ns. It is important to note that the ToF measurement does not require all bins to be evaluated. Instead, accurate ToF measurements can be extracted using a limited number of bins. The condition is that the laser pulse width must be longer than the pulse width of *enable*, as under this circumstance the laser pulse will fall inside two bins and the CoM algorithm can be effectively employed.

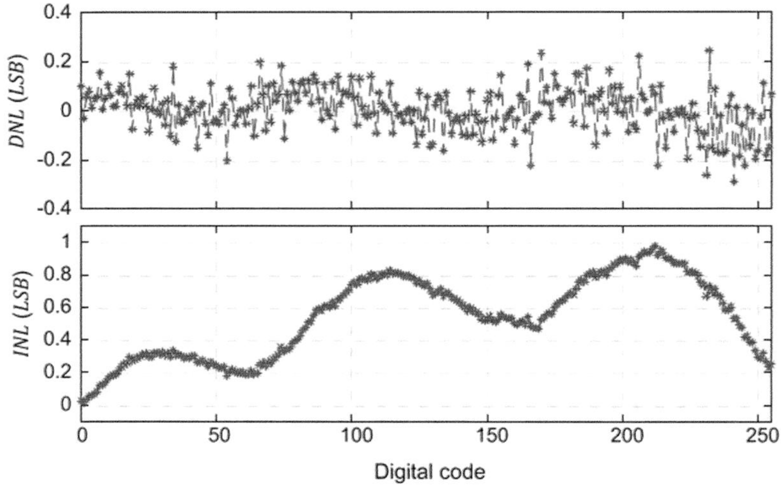

FIGURE 5.27 Differential and integral non-linearities (DNL and INL, respectively) of the programmable delay.

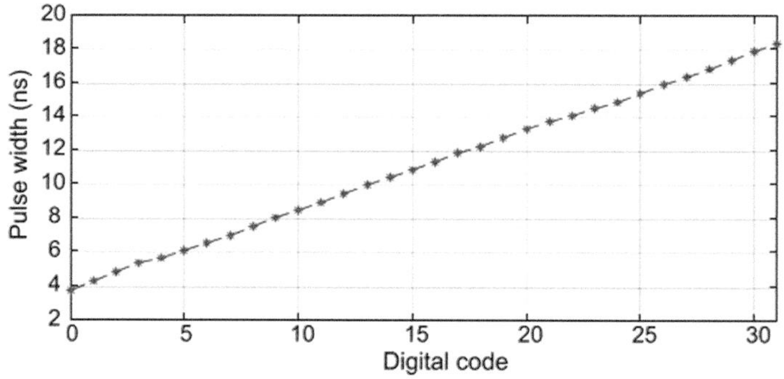

FIGURE 5.28 Pulse width of the enable signal as a function of the digital code.

Finally, temperature dependence of the programmable delay is attributed to the current reference and temperature coefficients of capacitors within the block. While the latter are implemented using metal-insulator-metal capacitors, their temperature coefficient is low. This results in a larger dependency on the reference current, whose impact will be discussed in Section 5.3.3.

5.3.3 BIASING CIRCUITRY

The performance of the biasing circuitry directly affects the rest of blocks in the sensor since any variation or noise will be coupled. Therefore, references must be insensitive to PVT variations for a proper operation. In particular, the SAER vision sensor requires a reference current independent of the temperature for the programmable delay and dead time of SPADs. The impact of the latter remains insignificant, provided that the dead time does not exceed its minimum allowable value. However, alterations in the programmable delay manifest as errors in the absolute depth.

The reference current ideally relies on two primary factors: the bandgap voltage and the resistance value. In reality, the offset of the amplifier is also combined with the bandgap voltage, resulting in inaccuracies in the absolute current. While the absolute current value might present a tolerance or undergo calibration after the manufacturing process, the dependence on temperature of the offset affects the measurement. As the SAER vision sensor was conceived as a proof-of-concept prototype, the specifications can be more flexible. Additionally, temperature dynamics are slow and are unlikely to affect a single measurement. However, for a prototype intended for real-world operation, canceling the offset of the amplifier becomes crucial, either through calibrating the amplifier or using auto-zero or chopper modulation techniques [Enz96].

The temperature dependency of the bandgap voltage is depicted in Figure 5.29 for five different samples. Variations in the absolute voltage value fall within the range of ±50 mV, aligning with the values observed in simulation during the design phase. Furthermore, the temperature coefficient ranges approximately between 50–100 ppm/ºC for most samples, with one sample reaching 251 ppm/ºC. Although these values might be unacceptable for precision references, trimming [Ge11] and curvature correction [Ge11; Wang18] could be employed to account for process variations and enhance accuracy. However, for simplicity, trimming was only implemented in the resistor of the reference current to also compensate for variations in the resistance value.

Figure 5.30 illustrates the variations in the reference current with respect to temperature. An external resistor can be utilized for precise reference, reducing the dependence on temperature when using a low-temperature-coefficient resistor. Yet, the temperature of the die can be measured from the current of the bandgap reference, which is mirrored and measured off-chip, enabling compensation for these variations by adjusting the resistance of the internal resistor. Although this calibration might

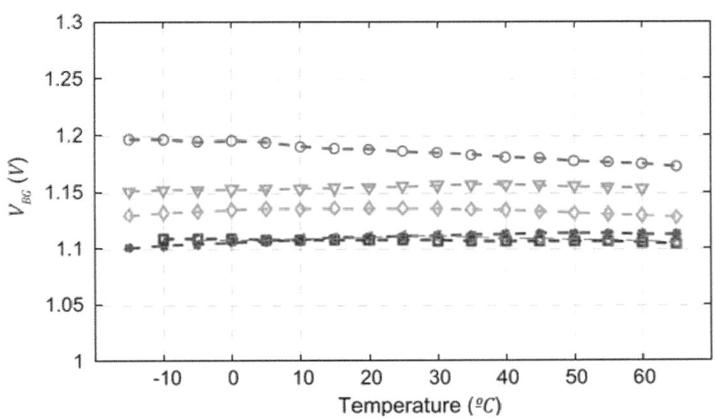

FIGURE 5.29 Bandgap voltage for five different samples.

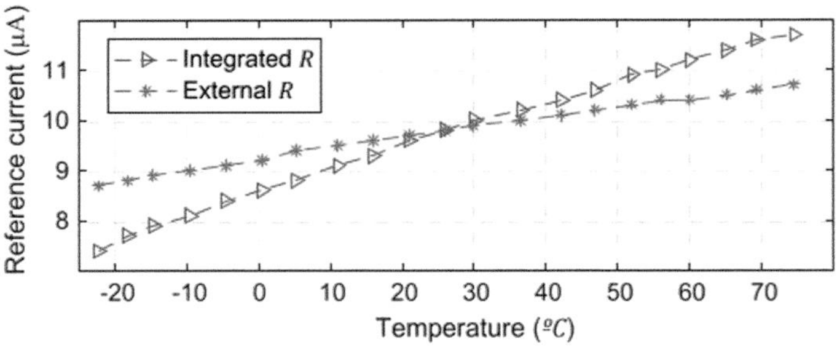

FIGURE 5.30 Reference current for programmable delay as a function of temperature.

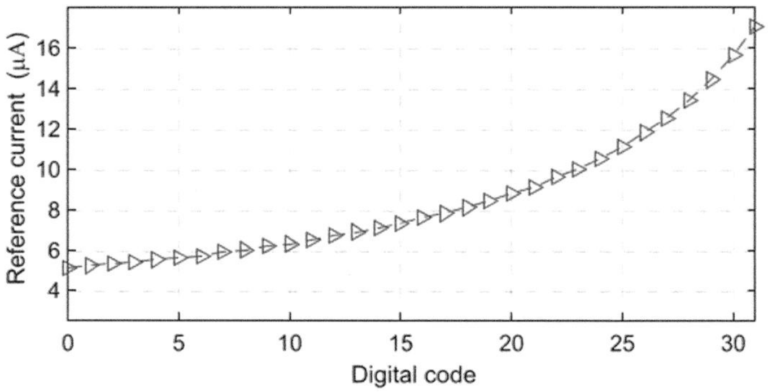

FIGURE 5.31 Reference current versus digital code.

not appear intuitive due to its nonlinearity, the ultimate goal is to achieve an accurate delay, which is inversely proportional to the current, allowing for a linear delay calibration if the resistor is trimmed.

Figure 5.32 (a) shows how the programmable delay varies with temperature. Although using an external resistor reduces deviations from ideal values, the use of a programmable internal resistor allows a quasi-lineal adjustment of this delay, as shown in Figure 5.32 (b).

Finally, Figure 5.33 displays the measured bandgap current employed to implement the temperature sensor. This allows estimating the temperature of the die with ±1°C accuracy and compensate for thermal effects during the test of the sensor.

5.3.4 Power consumption

The power consumption of the SAER vision can be divided into four different contributors.

- **Bias and references**. They present a maximum static power consumption of 2.06 mW, including bandgap, current reference, and programmable bias voltages.

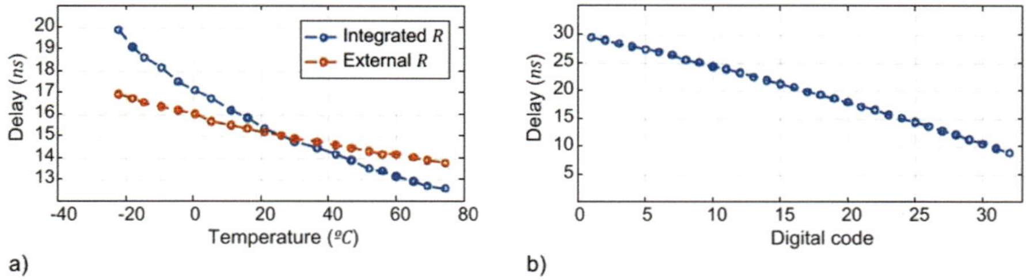

FIGURE 5.32 Temperature dependence of the programmable delay.

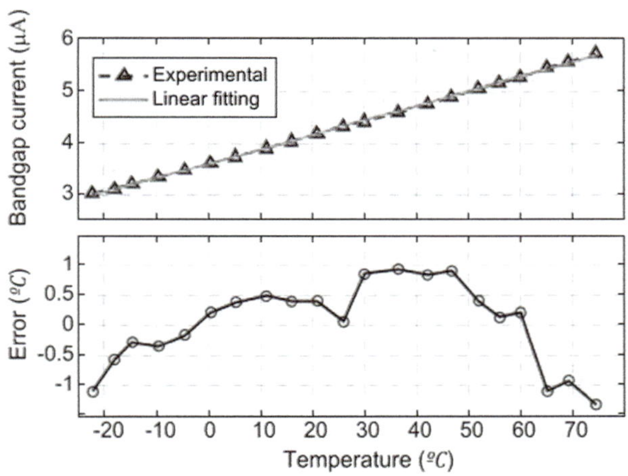

FIGURE 5.33 Output of the temperature sensor and error.

- **SPADs**. A large current flows through devices at the instant of photon detection.
- **Sensor and periphery**. Quenching and recharge, digital counters, and periphery blocks present static current due to leakage and dynamic current depending on activity.
- ***Input/Output* (IO) pads.** Transmitting information is an energy-intensive task due to the large package and PCB parasitics.

While bias and references account for static power consumption independent of the scene, power consumption related to SPAD, sensor, and IO significantly relies on the sensor's activity. Figure 5.34 illustrates the variation of each power consumption source concerning the event rate, when N_{ph} is set to 128. For a moderate event rate, i.e., lower than 100 keps, the power consumption of the pixel and peripheral circuitry dominates. However, as the event rate increases, the power consumption of SPADs and IOs starts to rise, becoming the main source of power consumption.

However, the event rate is directly reliant on N_{ph}. For a given scene, the event rate varies inversely with N_{ph}, causing a shift in the red and blue curves along the x-axis.

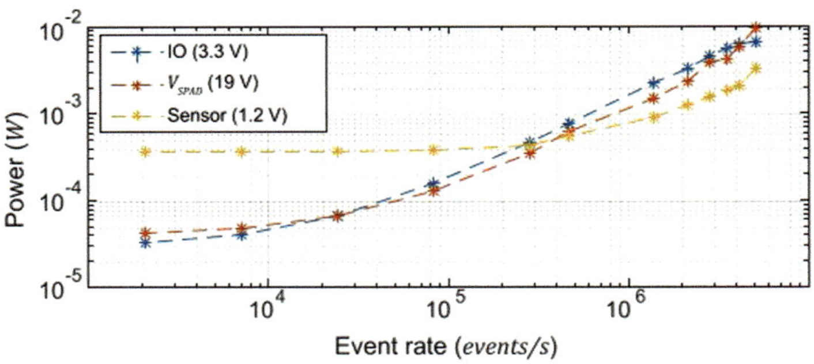

FIGURE 5.34 Power consumption of the SAER vision sensor.

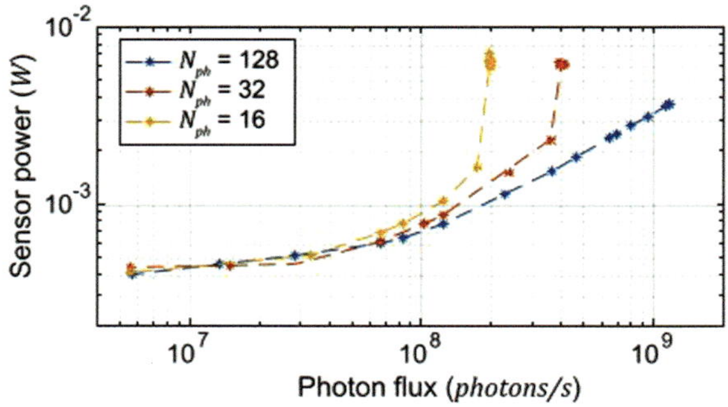

FIGURE 5.35 Power consumption of the sensor (pixel array and periphery) for different N_{ph} values.

Therefore, it is pertinent to illustrate the power consumption of the sensor (both the pixel array and periphery) concerning the total photon flux (evenly distributed across the entire array). Figure 5.35 depicts this power consumption across various N_{Ph} values.

At lower photon flux levels, power consumption remains consistent regardless of N_{Ph}. This suggests that the pixel circuitry's power consumption outweighs that of the periphery since the number of SPAD pulses remains constant, maintaining uniform activity within the array. Conversely, the event rate decreases with N_{ph}, leading to reduced periphery activity.

As the photon flux rises, higher $N - Ph$ values demonstrate decreased power consumption. This indicates that beyond approximately 1e8 photons/s, periphery power consumption begins to dominate. However, there is a sudden surge in power consumption beyond a certain flux, indicating saturation of the readout channel. During saturation, pull-up devices of unattended rows draw a static current consumption during the measurement. This state, though non-functional, serves as an upper limit for the event rate, which increases with N_{ph}. It is noteworthy that when N_{ph} equals 128, the sensor does not reach saturation within the measurement range.

5.4 SYSTEM CHARACTERIZATION

This section addresses a system-level characterization to assess the performance of the overall system. This evaluation encompasses the PMU to verify its capability in generating the SPAD voltage, assessing the ability of the optical emitter to generate a rapid and sharp pulse, and evaluating the functionality of the sensor during regular operation.

5.4.1 Power Management Unit

The PMU of the SAER vision sensor is composed of two main components, namely the DC-DC converter and the high-voltage LDO. While the former generates a high voltage from the 5 V input power supply, the latter regulates the output voltage.

Figure 5.36 shows the output of the DC-DC converter and the LDO during the startup of the PMU. While the DC-DC converter requires approximately 240 μs for the startup, the settling time of the LDO is 401 μs. Therefore, the time required for the PMU to activate the SPAD voltage is slightly lower than 1 ms. This metric is important if this voltage is supposed to be turned off after the measurement for an efficient operation during idle state.

Nevertheless, the most important characteristic of the PMU is its efficiency. Ideally, the PMU should exhibit a 100% efficiency, meaning that all the energy is transferred to the load. However, in practice, this is far from being true. Control logic and power losses in power devices such as the power MOS of the DC-DC converter reduce efficiency, which strongly depends on the load current.

Figure 5.37 depicts the efficiency of the PMU as a function of the load current. In this experiment V_{SPAD} was set to 20 V. Since the power consumption of SPADs was unknown and overestimated during the design phase, the PMU was not optimized for efficiency. In fact, the manufacturer of the DC-DC converter reports an efficiency peak at a current load of 50 mA, which is never reached under regular conditions.

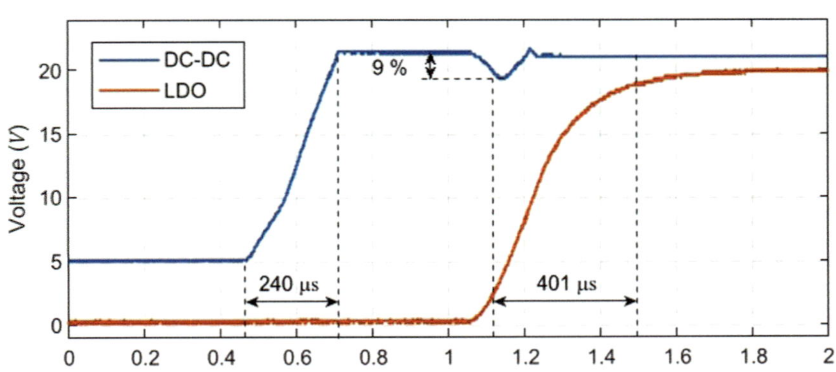

FIGURE 5.36 Start-up waveforms of power management unit.

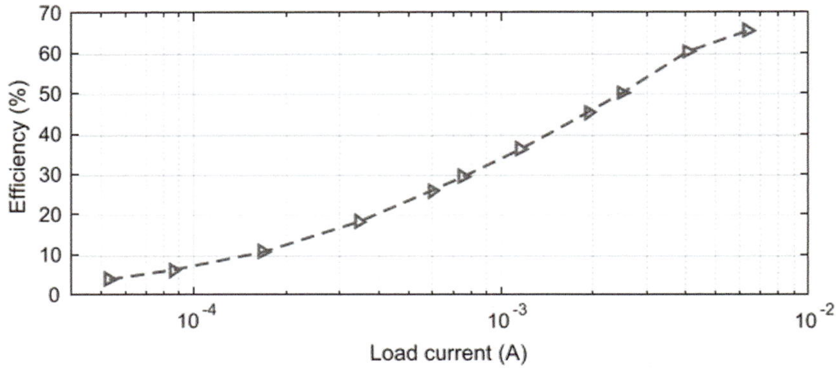

FIGURE 5.37 Efficiency of the PMU as a function of the load current @ $V_{SPAD} = 20$ V.

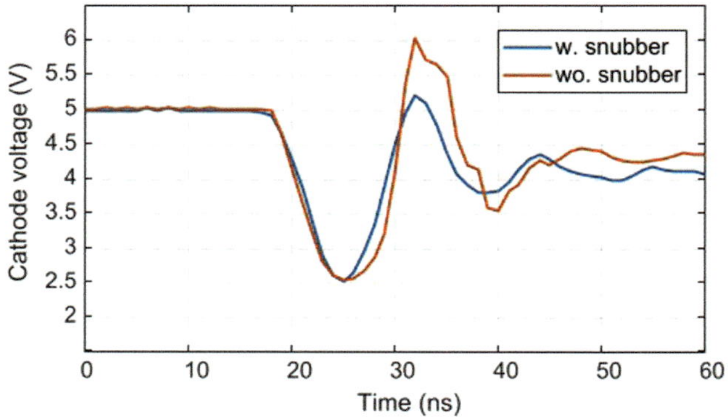

FIGURE 5.38 Cathode voltage of the laser with and without snubber.

Therefore, a redesign of the DC-DC converter is needed to optimize power consumption of the SAER vision system.

5.4.2 Optical Emitter

The optical emitter plays a crucial role in the ToF measurement. As explained in Section 2.2.1, a short, high-power and sharp laser pulse is desired to guarantee a proper operation. Hence, precise control of the laser is crucial. The SAER vision sensor incorporates two optical emitters: an EEL for long-range and a VCSEL for short-range scenarios. Although both devices exhibit similar specifications regarding rising and falling edges, handling higher currents is notably more challenging with the EEL, presenting a more complex scenario compared to the VCSEL.

As a first step in characterizing the laser, we monitored the cathode voltage of the EEL, directly linked to the driving current and hence, to the optical power. Figure 5.38 illustrates the cathode voltage during a 10 ns pulse. An independent experiment excluding the snubber highlights the importance of this circuit. Its absence resulted in observable risks of device damage caused by overshoot. Note that the cathode does not immediately return to its original level. This occurs because once the driver is turned off, the parasitic capacitance at the cathode is charged by means of the forward current. Thus, once the forward bias voltage falls below the threshold voltage of the diode, its dynamics slow down. However, this does not pose a concern, as emission is also minimal within this regime.

Figure 5.39 illustrates the laser pulse received from the sensor output. In practice, this pulse corresponds to the convolution of the enable signal and the laser pulse, thereby enlarging the pulse width by the width of the enable signal, set at 3 ns. Variations in results across pixels remain minimal. Figure 5.40 displays the rise time, fall time, and FWHM across the pixel array. Given the small number of pixels, gradients are barely discernible. Table 5.3 presents the mean values and standard deviations of these parameters.

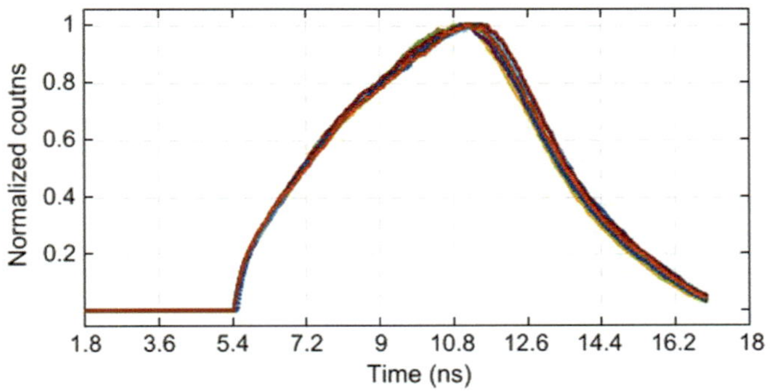

FIGURE 5.39 Laser pulse detected from the sensor.

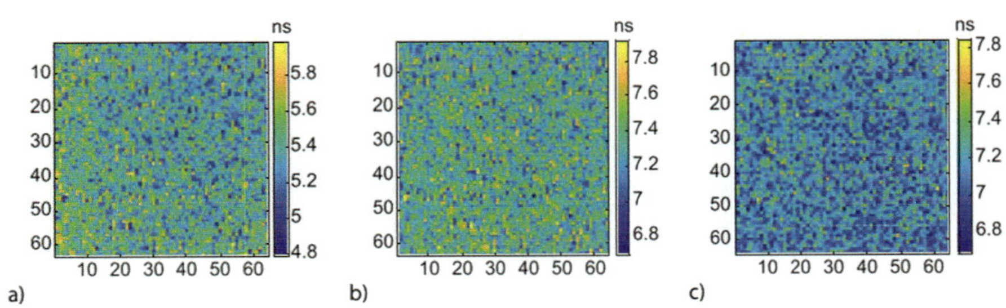

FIGURE 5.40 Parameters characterizing the received pulse across the pixel array: a) rise Time, b) fall Time, c) full width at half maximum.

TABLE 5.2
Parameters of the received laser pulse.

Parameter	Median	Standard deviation
Rise time	5.14 ns	196.1 ps
Fall time	7.33 ns	169.6 ps
FWHM	7.13 ns	160.7 ps

5.4.3 2D Imaging

In this section, we delve into the capability of the SAER vision sensor to acquire intensity representations. As outlined in Chapter 3, the sensor exhibits versatile operational modes, including the FR mode, TNP mode, and QI mode. Each of these modes presents distinct advantages and trade-offs, influencing the performance of the sensor in capturing 2D visual information.

Free-running mode

In FR mode, the SAER pixel functions as an I&F neuron. It continuously detects photons and generates an event when a tunable threshold of photons is reached. Consequently, the output pulse train of each pixel encodes the illumination level through its average frequency. This time-encoding method not only facilitates efficient data transmission to a neuromorphic processor but also offers an intra-scene dynamic range unrestricted by the depth of the counter.

The primary parameter of interest in this mode is the average frequency of the output concerning illumination. This measurement provides insights into both the system's dynamic range and the linearity of the measurement, considering any potential delays introduced during event generation and transmission.

To evaluate this functionality, we set up an experimental configuration with an Oriel 66884 halogen lamp that served as the light source, followed by three filter holders to enable precise light attenuation. The resulting light was directed to an integration sphere connected to the sensor, ensuring uniform illumination of the pixel array.

Figure 5.41 displays the relationship between the average frequency of output pulses and the illumination level. Data was collected by timestamping all generated events across the pixel array during a defined exposure period. For this experiment, all pixels were programmed with $N_{ph} = 128$. The lower limit is constrained by DCR, while the upper limit is determined by the dead time of the SPADs, which was set at

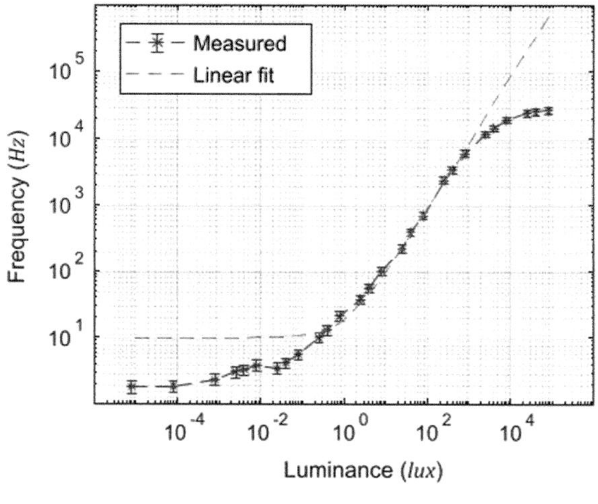

FIGURE 5.41 Spiking frequency as a function of the luminance.

1 μs. It is worth noting that the upper limit could be significantly extended if more than one SPAD was employed per pixel.

The sensor exhibits a linear response from approximately 0.1 lux to 1 klux without the use of attenuators. The detection range extends from nearly 0.01 lux to 10 klux, resulting in a linear dynamic range of 80 dB and a complete dynamic range of 120 dB.

Lastly, Figure 5.42 showcases an output image from the sensor in FR mode. While the sensor is not designed for frame reconstruction, this frame was constructed by enabling the sensor for a defined exposure time and measuring the event frequency for each pixel. In this case, the exposure time was 10 ms.

Time-to-N-Photons mode

The TNP mode is a specific case of the Octopus mode in which each pixel transmits only one event. Consequently, the illuminance vs. time behavior corresponds to that depicted in Figure 5.41. However, the SNR is constrained by N_{ph} and cannot be improved by averaging more than one event. Figure 5.43 demonstrates the SNR of

FIGURE 5.42 Sample image in Free-Running mode with 20,680 events received in 10 ms.

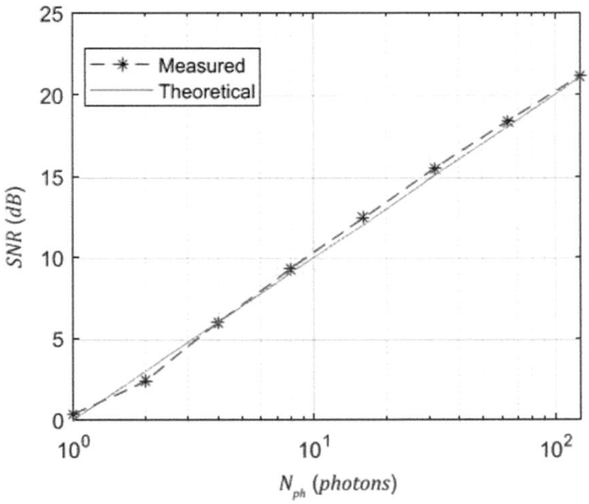

FIGURE 5.43 SNR of the measurement as a function of N_{ph} in the TNP mode.

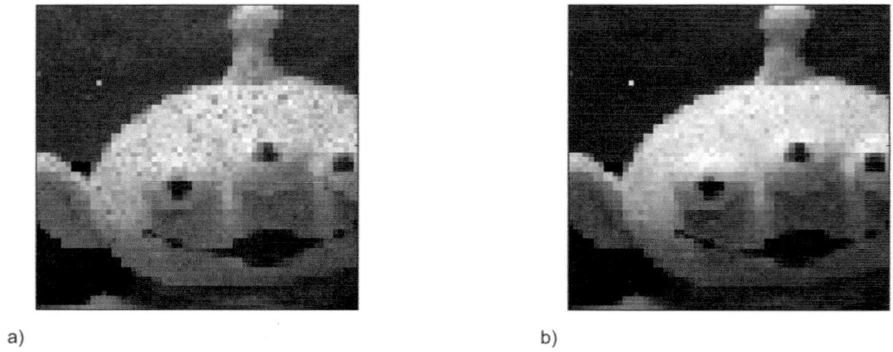

FIGURE 5.44 Rendered image in the: a) TNP mode, b) FR mode. Acquisition time is 32 ms for both.

TNP photons as a function of N_{ph}, aligning with theoretical expectations dictated by photon shot noise.

The primary advantage of the TNP mode lies in its reduced data throughput compared to the Octopus mode. While the operation of this mode is inherently synchronized with a signal that requests the measurement, each pixel is restricted to sending a single event. In this operating mode, highly illuminated pixels will not collapse the readout channel, but signals are more susceptible to distortion due to collisions, as measurements rely on a single event.

Finally, Figure 5.44 provides a comparative view of an image captured in both FR and TNP modes, utilizing the same exposure time. An interesting observation is that bright areas exhibit a higher SNR in Octopus mode due to the utilization of more than one event for image reconstruction.

Quanta Imaging mode

The SAER vision system introduces a multi-bit QI mode. While temporal information on events is lost and the operation is closer to the frame-based paradigm in this mode, it facilitates a continuous stream of 1-bit frames that can subsequently be integrated to generate 2D images.

The main advantage of the SAER vision sensor lies in the asynchronous readout of pixels detecting more than a defined photon threshold, denoted as k. This strategy improves readout speed, particularly in scenarios where the density of ones is reduced.

As presented in Chapter 3, there exists an option to reconfigure the in-pixel counter for the rudimentary integration of 1-bit frames. This entails setting N_{ph} to a value greater than 1, without resetting the counter before the subsequent exposure. Consequently, the number of events transmitted by pixels is diminished by a factor of N_{ph}, as exemplified in Figure 5.45, where dashed line represents $N_{ph} = 1$ and black line $N_{ph} = 4$.

Nonetheless, the utilization of a large value for N_{ph} may introduce significant uncertainty, as the counter may have a residue after all exposures. Therefore, if the

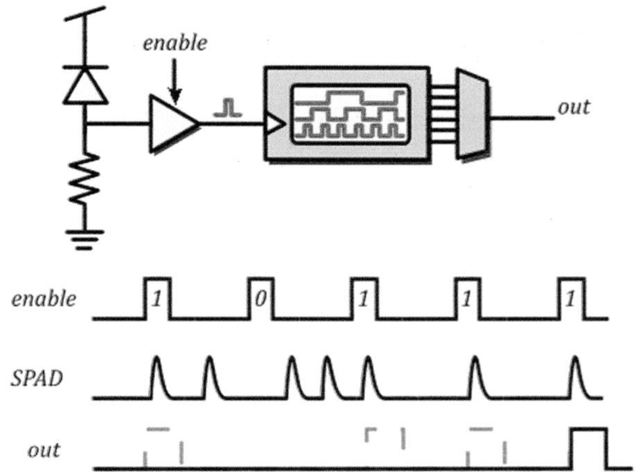

FIGURE 5.45 In-pixel pre-integration in the quanta imaging mode. Dashed line represents $N_{ph} = 1$ and black line $N_{ph} = 4$.

photon count of any pixel falls between 1 and N_{ph}-1 after the last sample, no event will be generated, causing uncertainty at those levels. This concern can be alleviated by dynamically reducing the value of N_{ph} as events are readout. Whenever an event is generated, the N_{ph} value for that specific pixel is adjusted to the immediately lower value than the number of remaining exposures. This mitigates the accumulation of information in the counter that will not be transmitted. However, it is important to note that in this approach, all counters start with the same value of N_{ph}. Thus, pixels with bit density lower than N_{ph}/N_{exp}, where N_{exp} is the number of exposures, will not transmit any information, posing an upper limit on the value of N_{ph}.

Figure 5.46 depicts the relationship between luminance and bit density in the reconstructed image, under an exposure time of 10 μs and for different k values. The results align with theoretical expectations [Foss13], affirming that the sensitivity of the pixels can be tailored through this parameter.

5.4.4 Depth computation

Figure 5.39 presented the sensor response to a laser pulse, displaying resulting histograms for various pixels. This experiment enabled ToF through computation of the CoM for each histogram, revealing a resulting non-uniformity of 0.7%.

To further evaluate sensor performance in ToF mode, distance sweeps were conducted with a moving target. Figure 5.47 illustrates the correlation between the target's distance and the sensor's output. The programmable delay covered a range of 22.53 ns with 32 steps of 704 picoseconds each. The measured range extended from 0.25 to 1.5 meters, constrained by the physical limitations of our experimental setup. Although the system exhibited 1.2% accuracy relative to the measured range, the operational range in this setup extends to 3.5 meters. Notably, the current source allows potential adjustments for distances up to 37 meters, restricted by the distances within the experimental setup and laser power.

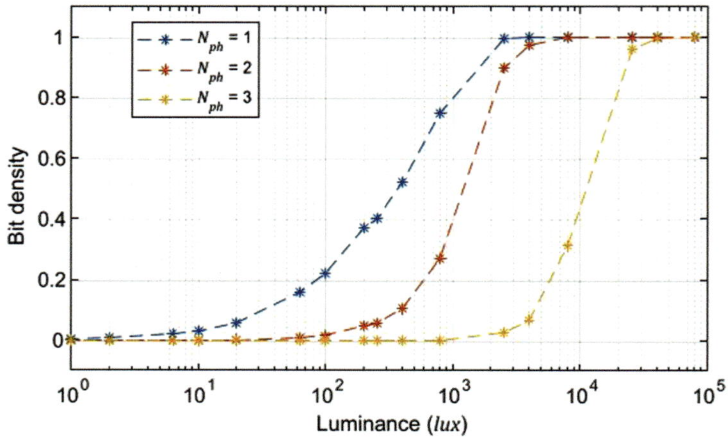

FIGURE 5.46 Output of the sensor in Quanta Imaging mode as a function of the luminance.

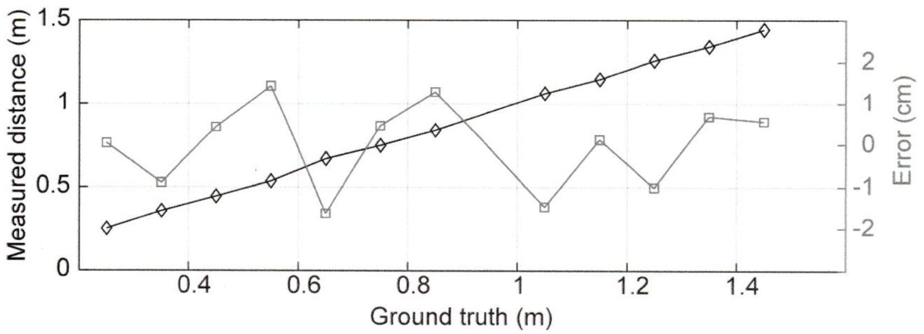

FIGURE 5.47 Comparison of measured distance to ground truth and resulting error.

The maximum error displayed in Figure 5.47 accounts for 1.5 cm. However, it is pertinent to highlight that the accuracy is constrained by the SNR of the histograms. This limitation is directly influenced by the number of counts, exposures, and laser power. Specifically, the PDP at NIR wavelengths in this setup falls below 3%. Such a scenario requires a considerable number of exposures to achieve a discernible histogram peak, consequently imposing significant limitations on the achievable frame rate. Therefore, customized device engineering becomes imperative to enhance and optimize ToF measurements. Moreover, an effective approach towards improving SNR involves the integration of multiple SPADs per pixel alongside a combining network.

Figure 5.48 shows the histogram reconstructed from a ToF measurement using different exposures per bin. As seen, there exists an enormous difference between taking 1,000 or 10,000 samples per bin. However, this is not practical at all as it would require a long time for the measurement. Indeed, measurement can be extracted from a reduced number of bins.

Figure 5.49 shows an example of an intensity and a depth image acquired by the sensor, along with reference images from a standard camera. In this example, 32 bins

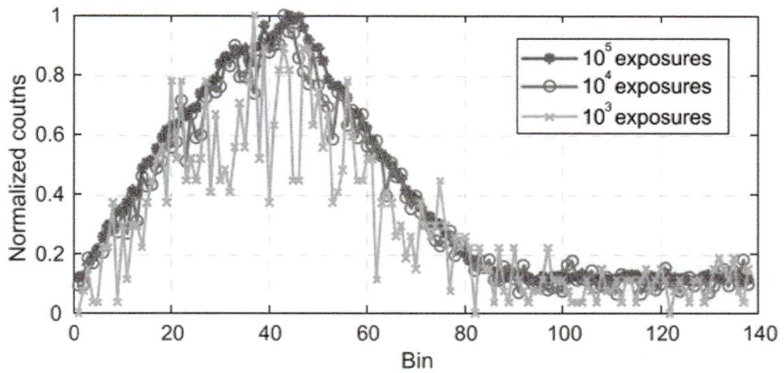

FIGURE 5.48 ToF Histogram built from different exposures per bin.

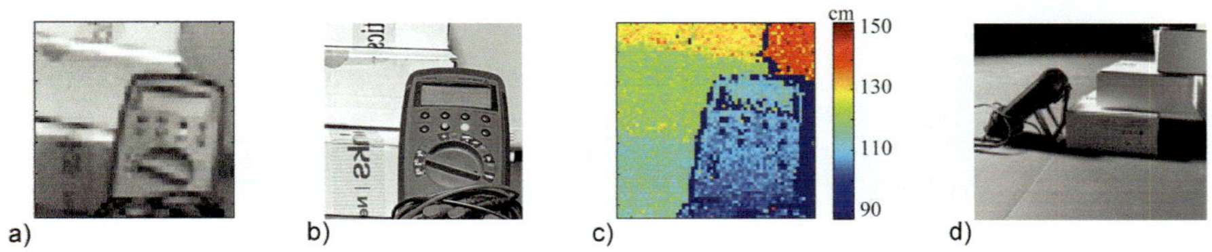

FIGURE 5.49 a) Intensity map acquired by the sensor. b) Image from reference camera. c) Depth map measured with the sensor. d) Side image for scene depth reference.

were swept to compute ToF. Pixels with a SNR lower than 10 were not considered in the computation and were labeled as the minimum distance (dark blue).

Reconstructing depth images is feasible even with a reduced number of bins. Figure 5.50 displays the reconstructed image utilizing 32, 16, 8, and 4 bins. However, reducing the number of bins results in reduced accuracy and loss of image details. This decline primarily stems from the non-ideal laser pulse, as evident in Figure 5.48, showcasing a reconstructed pulse with prolonged transients rather than a sharp rise and fall sequence. Consequently, utilizing fewer points to construct the histogram leads to significant signal distortion. In contrast, if the histogram was a squared pulse, the signal would be less affected by which particular bin is being used for sampling.

To enhance the SNR of the measurement, pixels can be grouped into clusters of 2×2 pixels, forming macropixels [Vorn20]. This setup allows the counting of more than one photon per exposure, but it reduces the resolution of the depth map. Essentially, this process just involves disregarding the LSB of the AER bus during the readout of events.

Figure 5.51 displays the depth map reconstructed using a 32×32 pixel array. In this scenario, a greater number of pixels meet the minimum SNR requirement for accurate measurement, resulting in improved uniformity. However, this improvement comes at the cost of reduced spatial resolution. It is noteworthy that achieving a similar effect could be possible by incorporating four SPADs within each pixel.

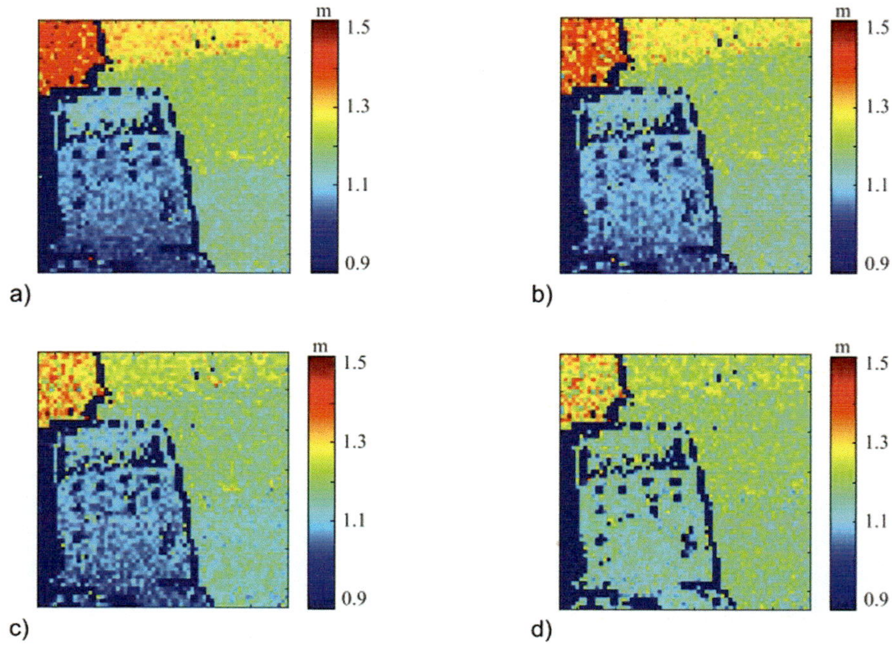

FIGURE 5.50 Depth map reconstructed using: a) 32 bins, b) 16 bins, c) 8 bins, d) 4 bins.

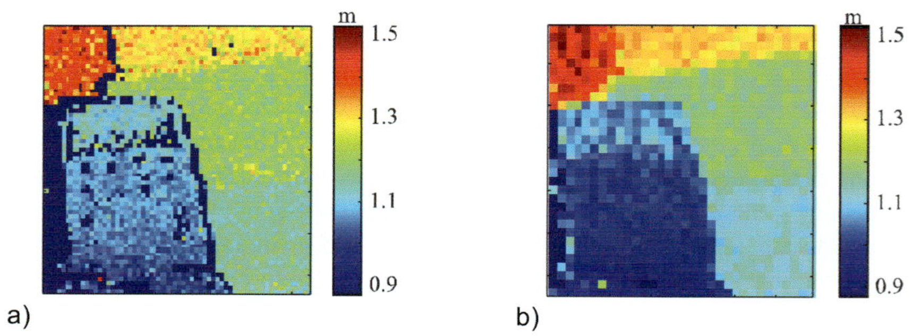

FIGURE 5.51 a) Original depth image. b) Downscaled depth image using 2x2 pixels as a macropixel.

5.4.5 Discrete Dynamic Vision

To validate the theory described in Section 2.3.3, raw event data was collected from a pixel operating in FR mode under different luminance conditions. This raw dataset comprises inter-arrival times of events, each corresponding to the detection of $N_{ph} = 128$ photons. This dataset serves as the basis for estimating the probability of event detection, defined as the likelihood of the relative difference between two inter-arrival times exceeding a specified threshold.

Initially, the probability of making a wrong decision was estimated by computing $\hat{\alpha}$ using each value of the dataset as T_{rst} and detecting how many samples exceeded α_{th}. Additionally, the analysis was repeated using the inter-arrival times of 4 and 16 consecutive samples to virtually increase N_{ph} to 512 and 1024, respectively. The experimental data for these scenarios is illustrated in Figure 5.52, aligning closely with the theoretical findings detailed in Section 2.3.3. It is important to note that the probability of false detection relies solely on α_{th} and N_{ph}, rather than the absolute luminance value. However, since comparisons are carried out upon event detection, the rate of false detections depends on the illumination. Furthermore, Figure 5.52 shows that achieving a contrast sensitivity below 5% with minimal noise is feasible; however, it requires an increase in the number of photons, consequently leading to increased latency.

Following this, a parallel experiment was executed, involving the calculation of $\hat{\alpha}$ utilizing data from different luminance levels. The actual α was determined by computing the average value from each dataset. Figure 5.53 illustrates the experimental outcomes, which demonstrate a strong alignment with the theoretical model. When defining contrast sensitivity as the point where P_{det} reaches 0.5, it becomes apparent that this metric is solely reliant on α_{th} and aligns precisely with its value.

To complete the validation of the model outlined in Section 2.3.3, an analysis was conducted to determine the necessary number of consecutive samples required to detect a variation, assuming an ideal transition from one dataset to another. By utilizing each sample from the reference dataset as T_{rst}, the number of samples required to trigger an event was extracted. This procedure facilitated the establishment of an average response time for each value of α. Figure 5.54 illustrates this relationship relative to the inter-arrival time of the new intensity level, T_2. It is important to note that, as indicated in the theoretical model, decreasing N_{ph} can enhance the response time. Furthermore, reducing α_{th} can amplify P_{det}, yet such adjustments can elevate the probability of false detections.

One noteworthy aspect of implementing a discrete DVS in the digital domain is its uniformity. Processing in this domain eliminates susceptibility to mismatches or noise in the signal. Thus, the only source of non-uniformity in operation lies within the SPAD front-end. While sensitivity mismatches do not affect temporal contrast

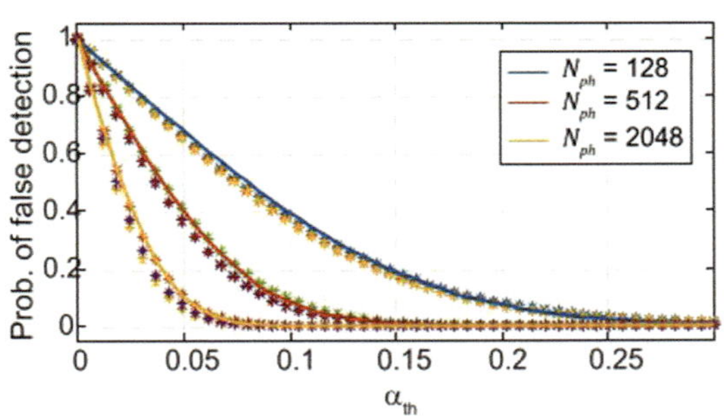

FIGURE 5.52 Probability of false detection in the discrete DVS mode as a function of α_{th}.

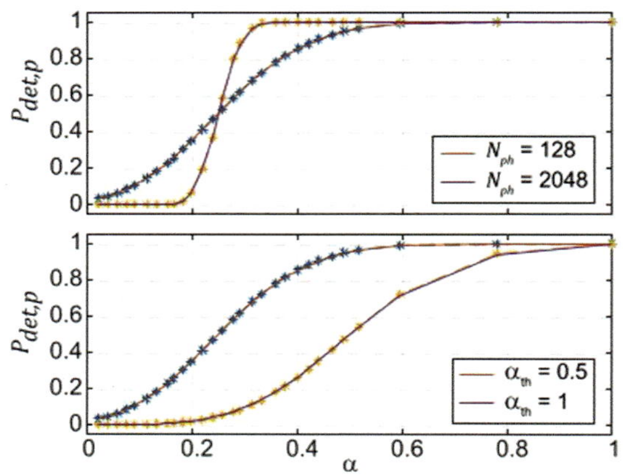

FIGURE 5.53 Probability of detection in the discrete DVS mode as a function of N_{ph} and α_{th}.

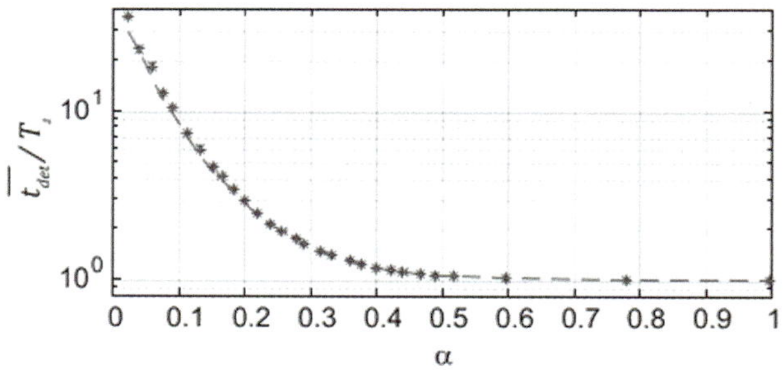

FIGURE 5.54 Latency relative to T_2 as a function of α with $\alpha_{th} = 0.25$ and $N_{ph} = 128$.

computation, discrepancies in DCR may introduce non-uniformities. However, as depicted in Figure 5.4, the median DCR remains below 10 Hz, rendering it insignificant under moderate intensity conditions. Nevertheless, hot pixels fail to detect variations as the signal is hindered by DCR.

Although the primary metrics of a discrete DVS are well defined by the model outlined in Section 2.3.3, the sensor's response can be limited by the photon collection process, specifically by the SPAD and photon integrator, which filter intensity variations beyond the detector's bandwidth. To study the sensor response, a laser was excited with a square wave signal exhibiting positive intensity variations of α = 0.5. The reason behind employing a square wave signal is based on the expectation of generating a number of events per cycle equal to one. Figure 5.55 shows the experimental results as a function of the frequency of the laser signal, for different values of α_{th} and N_{ph}. Only positive events are shown since results barely vary for the case of negative events. The results indicate that the sensor's bandwidth is restricted to approximately 30 Hz to 1 kHz, depending on the configuration of α_{th} and N_{ph}. Nevertheless, it should be noted that as the comparison is triggered by the collection

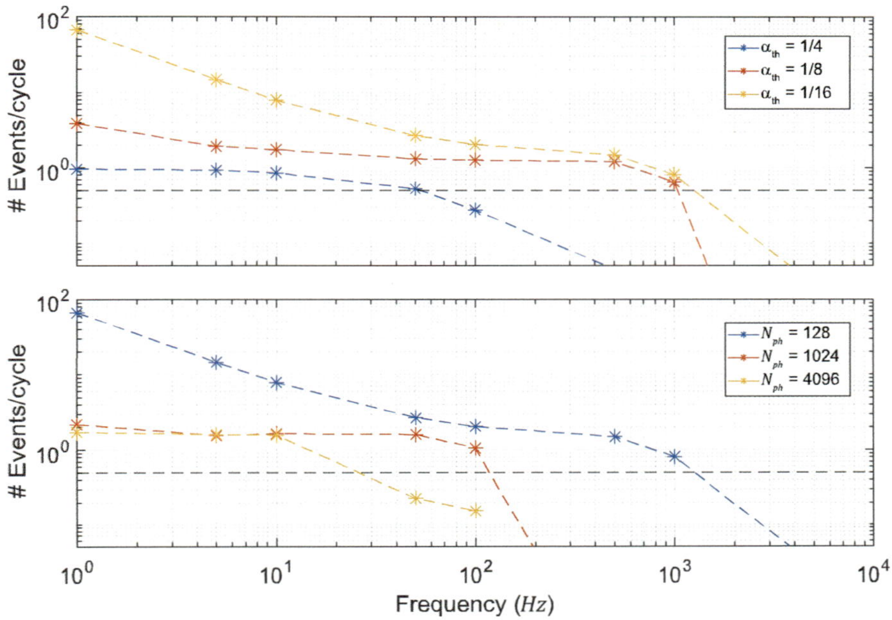

FIGURE 5.55 Sensor response observed under a square wave with positive steps of $\alpha = 0.5$ Top: Maintaining a fixed value of $N_{ph} = 128$. Bottom: Using a fixed $\alpha_{th} = 1/16$. The intersection with the black line (0.5 eps) represents the bandwidth of the sensor.

of N_{ph} photons, this bandwidth depends linearly on the intensity of the stimuli. In fact, these parameters could be dynamically adjusted if aiming for a constant bandwidth, at the cost of an intensity-dependence probability of false detections.

At lower frequencies, when either α_{th} or N_{ph} is too low, the number of events exceeds one. This occurs because a longer signal period entails a higher count of intensity events, resulting in more comparisons. Consequently, the number of spurious events diminishes as the frequency increases, given a constant intensity. In contrast, the number of events per cycle might fall below one at high frequencies, leading to signal averaging and a reduction in step amplitude. Consequently, the count of events per cycle experiences a sudden drop. Once again, a trade-off emerges between speed (photoreceptor bandwidth) and accuracy (contrast sensitivity).

Then, the event classifier was configured to tag events related to motion detection. Four subsequent events of the same pixel were used to trigger the event classifier, virtually increasing N_{ph} to 512 photons. Additionally, α_{th} was configured as 1/4. Consequently, the system output comprises a series of events, each accompanied by an address, timestamp, and sign. Figure 5.56 represents positive and negative events as a point cloud that evolves over time. At the beginning of the recording, a fan is activated, progressively increasing its rotation speed. In this scenario, the fan constitutes the sole moving object, except for sporadic points reflecting the user's interaction with the fan.

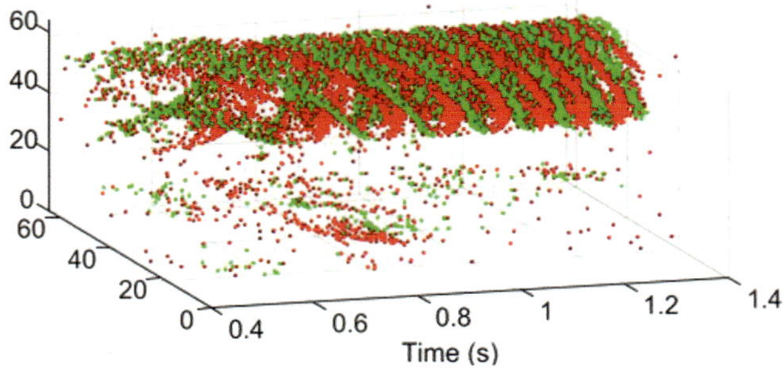

FIGURE 5.56 Point cloud corresponding to positive (green) and negative (red) events. The scene shows the transition of a rotating fan from off to on state.

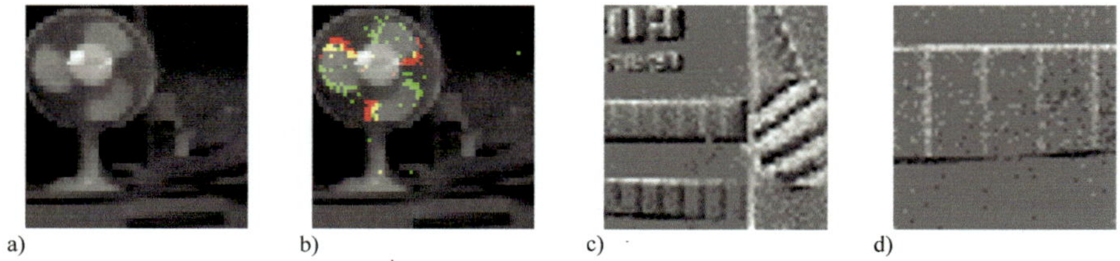

FIGURE 5.57 a) Image rendered from intensity events. b) Motion events overlapping intensity image. c) Edmund density target with a calibrated contrast of 0.1. d) Detail of Edmund density target.

Finally, Figure 5.57 (a) shows the reconstructed image using intensity events after an exposure time of 10 ms. A total of 20,608 events were attended. Figure 5.57 (b) depicts the same image highlighting pixels that detected motion within the given time frame, demonstrating a reduced event count with only 266 events. This includes both the motion-related event and the subsequent event used to compute the intensity level. In addition to this scene, Figure 5.57 (c) and (d) show the response of the sensor under a moving Edmund density target with a calibrated contrast of 0.1.

5.4.6 Adaptive sensing

To evaluate the effectiveness of dynamic in-pixel thresholding, we conducted an experiment in which the illumination of the scene was varied. The main goal of this algorithm is to keep a constant event rate of pixels by adjusting the value of N_{ph} individually, as described in Section 2.5.2.

Figure 5.58 shows the evolution of the sensor event rate over time. At $t = 0.33$ s, the light was turned off, causing the event rate to drop below 199 keps, thus wasting the readout channel bandwidth. However, after 0.1 s, the event rate returned to around 5.27 Meps, resulting in a reconstructed image that exhibits more detail. At $t = 0.87$ s, the light was turned on again, causing the readout channel to saturate due to a high event rate. Upon detecting this condition, the receiver resets the threshold for all pixels, initiating the adaptive process once more, demonstrating the ability of

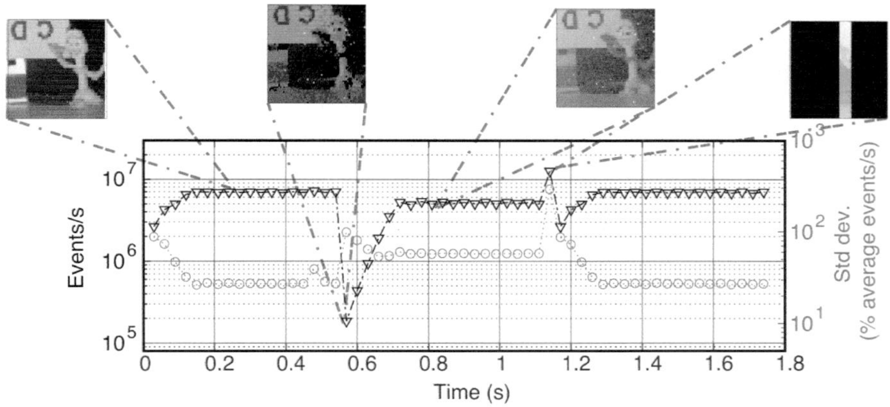

FIGURE 5.58 Evolution of the sensor event rate using the adaptive algorithm.

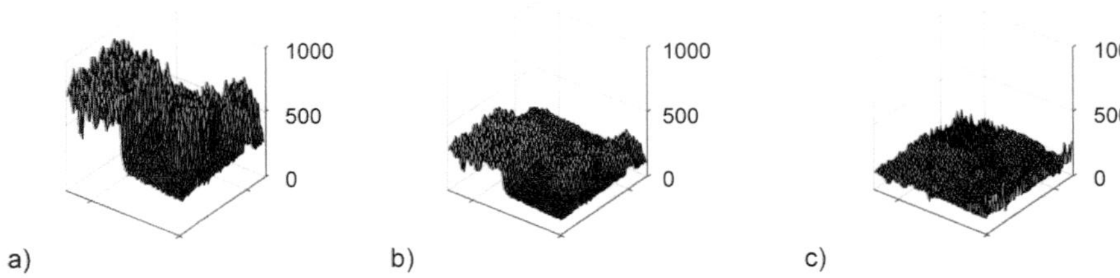

FIGURE 5.59 a) Event-rate map at t = 0 s (2.17 Meps). b) Event-rate map at t = 0.03 s (0.9 Meps). c) Event-rate map at t = 0.06 s (0.51 Meps).

the proposed sensor to adapt to scene illumination variations and optimize readout channel bandwidth consumption.

Figure 5.59 shows that the event rate of each pixel is equalized to ensure that all pixels consume the same bandwidth of the readout channel, which is reflected in the uniform distribution of event rates across the event-rate maps once the algorithm settles. Although the same number of photons must be collected to gather enough statistical information, this method guarantees the extraction of information with the maximum temporal resolution and to leverage the bandwidth of the readout channel.

5.4.7 Summary and Benchmarking

The validation and characterization of the camera system in controlled laboratory conditions allow for a comprehensive summary and comparison of its key attributes and performance against established methods. Table 5.3 provides an overview of the SAER vision system's performance and benchmarks it against existing state-of-the-art systems. It is worth pointing out that while the sensor serves as a proof of concept of the possibility of merging intensity and depth measurements with the discrete DVS

TABLE 5.3
State-of-the-art comparison.

Parameter	This work	[Rocc20]	[Gyon23]	[Guo23a]	[Ding23]
Functionality	Intensity, DVS and depth	Intensity, DVS and depth	Intensity, DVS and depth	Intensity and DVS	Intensity and DVS
Vision type	Discrete DVS-AER	Frame differencing	Frame differencing (depth)	Analog DVS	Scan-AER
Resolution	64 × 64	128 × 128	64 × 32	1032 × 928	128 × 128
Technology (nm)	110	40	40	40 BSI CIS +65 CMOS+ 40	180
Detector	SPAD	SPAD	SPAD	CIS	CIS
Local sensitivity control	Yes	No	No	No	No
Readout	Event-driven	Frame-based	Frame-based	Frame-based	Event-driven
Pixel pitch (μm)	24.5 × 24.5[(1)]	40 × 20	114 × 54	8×8 × 8×8	16.5 × 16.5
Fill factor (%)	3.5	13	16.3	6.25	5
Latency (DVS)	128 μs [(2)]	2 ms[(3)]	20 ms[(3)]	> 100 μs	-
Maximum frame rate (fps)	4070	500	50	18–120	> 1.16k
Dynamic range	> 100 dB	69 dB	> 120 dB	72 dB	105 dB
Min. contrast sensitivity (%)	< 5[(4)]	< 5[(4)]	-	15	-
3D imaging mechanism	dToF	dToF	dToF	None	None
Sensor's power consumption (mW)	3.4 @ 1 Meps[(5)]	55	70	64 @ 300 Meps	800
Normalized power consumption[(6)] (nJ)	4.3-19.6	6.7-22.5	683.59	0.56	42
ToF range (m)	0.25 (min) 37 (max) 1.5 (meas.)	1.5 (min) 48 (max) 3 (meas.)	50 (max)	-	-
ToF accuracy (% range)	1.2[(7)]	0.6	0.1	-	-

(1): Pixel does not include the event classifier in this implementation.
(2): Minimum time required to collect N_{ph} = 128 photons, limited by dead time of SPADs (1 μs).
(3): Normalized power consumption = Power/Frame rate/No. of pixels.
(4): Achievable if the number of photons is large enough.
(5): This metric does not include power consumption of the event classifier in DVS mode.
(6): Normalized to a 64x64 resolution and an equivalent frame rate of 20 fps.
(7): Accuracy results 0.45% of the range considering the entire range.

concept, optimization of the different sections is required to meet in-the-field application challenges.

Section 2.6 delved into the definition and exploration of limitations on different metrics in the context of the SAER vision sensor, such as dynamic range and equivalent frame rate. The sensor's dynamic range, capped by the DCR and the SPADs' dead time of 1 μs, showcases a median DCR below 10 Hz. This aligns with the 120 dB demonstrated in Section 5.4.3, although the linear dynamic range is currently restricted to 80 dB. However, this limitation is inherent to the SPAD device itself, suggesting potential enhancements. Comparing frame rates presents challenges, given that the concept of frames is no longer of interest in an event-driven sensor. Even so, an equivalent frame rate is a critical metric to establish. Section 2.6.2 defined a method to determine the maximum equivalent frame rate based on the readout channel's bandwidth and pixel count. Figure 5.60 illustrates the maximum dynamic range versus equivalent frame rate, constrained by the system's bandwidth. Notably, the maximum dynamic range is also limited by the bit-depth of the in-pixel counter. A one-bit increase in counter depth corresponds to a 6.02 dB rise in the maximum dynamic range. Moreover, as the equivalent frame rate depends on pixel count, scalability concerns arise, prompting the need for sophisticated techniques to overcome AER readout channel limitations, like AER grouping [Son17], event pipelining [Fina20] or skip-logic scanning [Guo23a].

In essence, the SAER vision sensor introduces an innovative approach to 3D and 2D imaging, taking advantage of the benefits of event-driven architectures where prioritizing information over data is fundamental. Through enhancements in pixel optimization and readout channel strategies, metrics like dynamic range, equivalent frame rate, and latency in the discrete DVS mode could potentially match or exceed the performance of established systems.

5.5 CASE STUDY: THE STELLAR OCCULTATION OF BETELGEUSE

During the pre-dawn hours of 12 December 2023, the luminous star α Orionis (Betelgeuse) was anticipated to undergo an occultation by the asteroid (319) Leona

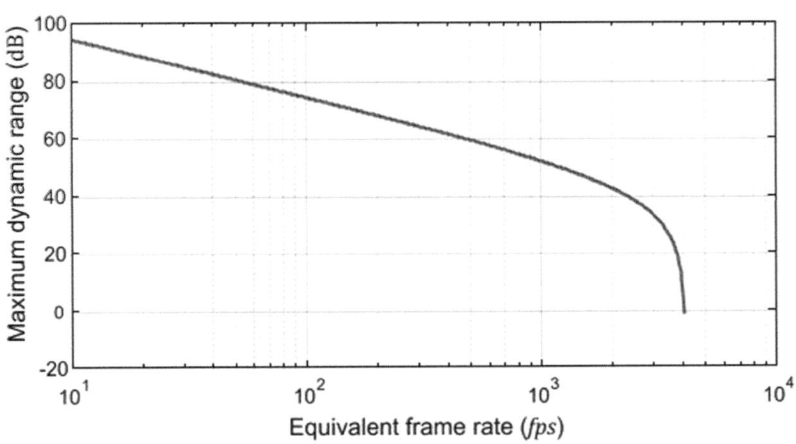

FIGURE 5.60 Maximum dynamic range as a function of the equivalent frame rate.

[Orti23]. The forecasted shadow path of this event, i.e., the trajectory on Earth where the occultation falls, was expected to cross a very narrow region over southern Spain, presenting an opportunity for observation of this extraordinary event.

This occultation stood as a unique event for astronomical research, offering a remarkable opportunity to precisely measure Betelgeuse's diameter and brightness distribution by observing the light curve as the asteroid progressively occulted the star. Unlike most stellar occultations, where stars' diameters are minuscule compared to the passing solar system minor bodies, Betelgeuse's immense angular diameter promised a distinctive occurrence of partial or total eclipse, provided Leona's angular diameter is sufficiently large. Moreover, the heightened interest in Betelgeuse, attributed to its recent significant dimming, had spurred diverse speculation, attributed to a dusty veil [Mont21]. This exceptional event provided an opportunity to test the performance of our sensor for astronomical observations.

The experiment was also prompted by the fact that the shadow path indicated the occultation would be observable near Sierra de Segura, Spain. The town of Nerpio, situated in this vicinity, is home to the astronomical facilities of AstroCamp [Astr], where the experiment took place. Figure 5.61 shows the experimental setup. The sensor was equipped in a 10-inch Ritchey-Chrétien telescope with a 2,000 mm focal length, designed to be remotely operated and controlled. The telescope-camera system provides a resolution of 2.53 arcseconds per pixel, covering a field of view of 2.4 arcmin × 2.4 arcmin.

Figure 5.62 shows an image of the constellation of Orion taken with a mobile phone camera on 12 December 2023, prior to the occultation. Betelgeuse is the second-brightest star in Orion and the tenth-brightest star in the night sky. This is noteworthy as a significant photon flux enables the acquisition of information at a high rate while maintaining a substantial SNR.

In astronomy, the accurate representation of images is not as crucial as preserving information. We intentionally defocused the image of Betelgeuse to distribute the photon flux across a broader area on the sensor. This method ensures the light curve of the star can be accurately measured while maintaining the SNR during the

FIGURE 5.61 Camera system equipped in the telescope.

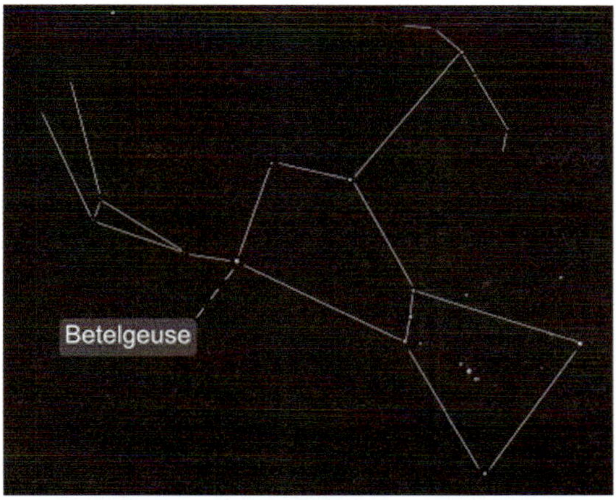

FIGURE 5.62 Image taken with a mobile phone camera highlighting Betelgeuse in constellation of Orion prior to the occultation.

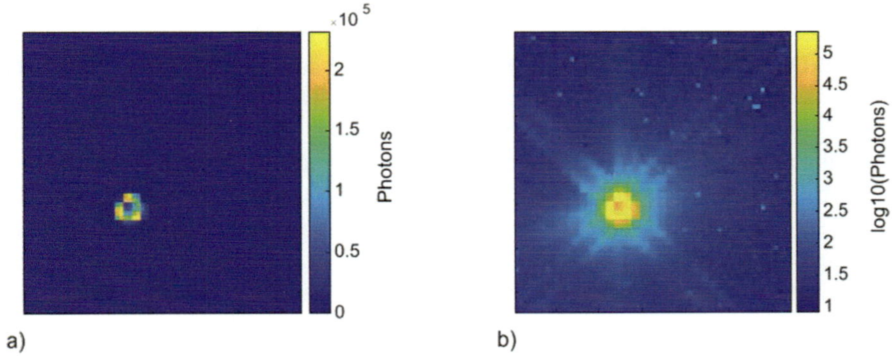

FIGURE 5.63 Image taken with the SAER vision sensor after 60s acquisition: a) linear and b) logarithmic scales.

observation. In the case of the SAER vision sensor, this approach proves advantageous as it prevents SPAD devices from becoming saturated and ensures the sensor's proper functionality by reducing the maximum photon flux to guarantee that the sensor operates in the linear regime. Figure 5.63 displays an image captured by the sensor prior to the occultation, utilizing events recorded in 60 s and correcting hot pixels with previous DCR data. The image illustrates a toroidal shape of Betelgeuse after defocusing the telescope. Figure 5.64 depicts the intensity profile in both axes, showcasing a FWHM of ~6 pixels in each direction, while 90% of the signal is contained within a circle that has an 8-pixel diameter.

The occultation was recorded by setting the number of photons required to trigger an event, N_{ph}, to 1 across all pixels to extract information with the highest temporal resolution possible. This configuration ensured that neither the readout circuitry nor

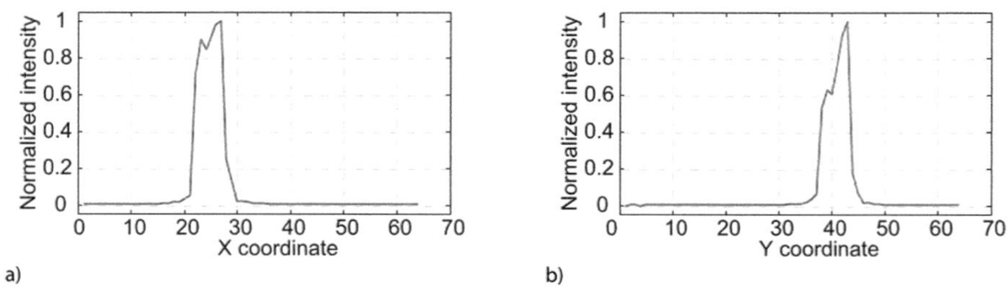

FIGURE 5.64 Intensity profile of Betelgeuse in a) X axis and b) Y axis.

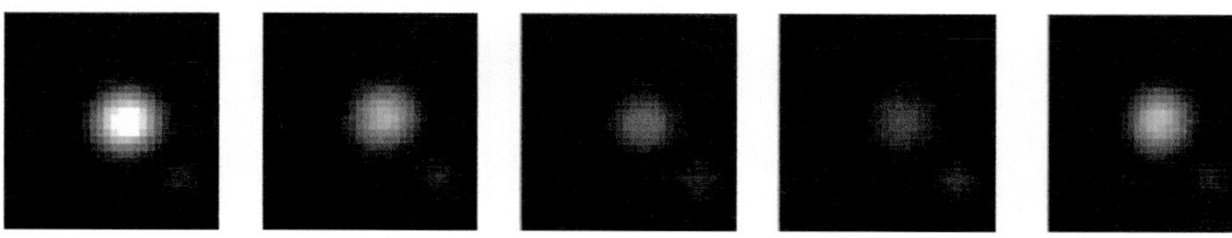

FIGURE 5.65 Illustration of Betelgeuse's eclipse through frames generated from event data.

the SPADs became saturated, preventing any signal corruption. The output of the sensor during this experiment comprises a list of events associated with the detection of individual photons, storing both the timestamp and the address of the pixel detecting the photon. Figure 5.66 depicts Betelgeuse's eclipse through several frames reconstructed from event data. A Gaussian filter was utilized to compensate for defocusing effects. Additionally, the small spot visible in the bottom right corner of the image corresponds to hot pixels.

However, in astronomy, frames hold minimal interest. Instead, the ROI can be regarded as a macropixel where the signal is spatially integrated. Analyzing variations in total photon flux over time facilitates the extraction of a light curve. A light curve provides a graphical representation of a star's brightness variations over a specific time span. These fluctuations can arise from diverse factors, encompassing intrinsic stellar attributes like pulsation, rotation, or stellar activity, as well as extrinsic elements such as eclipses. Astronomers use light curves to discern patterns, periodicities, and irregularities, extracting crucial information about the star's behavior, structure, and surrounding environment. The analysis of light curves serves a crucial role in uncovering celestial phenomena, unraveling the nature of stars, and detecting significant astronomical events.

The integrated signal can be derived by measuring the inter-arrival time of a specific number of photons, revealing a trade-off between temporal resolution and photon shot noise. Figure 5.66 displays the light curve obtained with the sensor using 4 and 1,000 photons as an event to form the signal, along with its moving average represented in red color. The temporal axis is relative to the beginning of the observation. The event rate determines the time step of each point on the curve, demonstrating a temporal resolution at the microsecond level—about four orders of

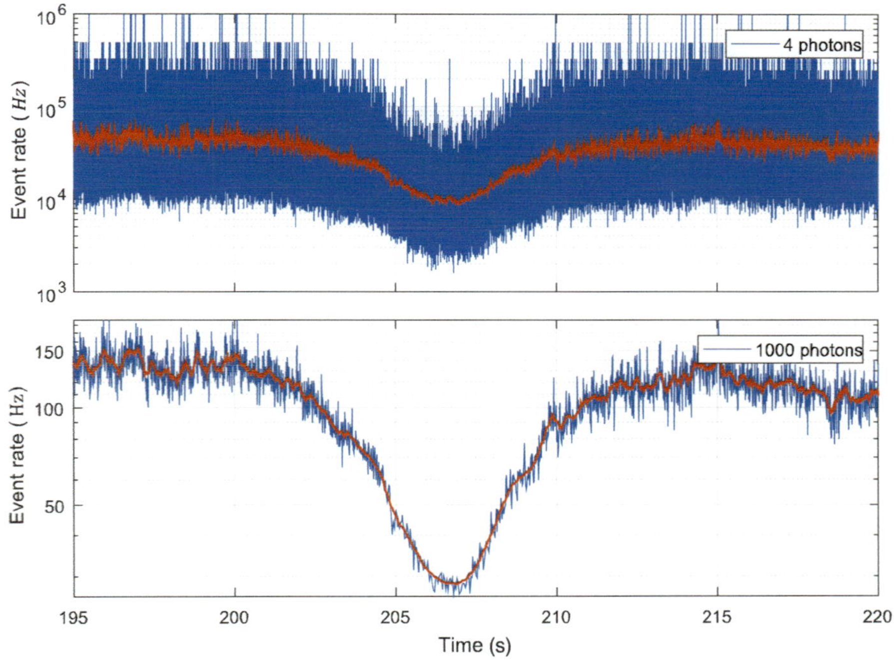

FIGURE 5.66 Light curve of the occultation of Betelgeuse using 4 and 1000 photons per sample. Red lines represent the moving average.

magnitude superior to CMOS and CCD (typical exposure times of 50 ms were used for this eclipse), which also have readout noise that deteriorates the SNR.

It is important to note that a higher rate and SNR could be achieved with a larger photon collection. Therefore, enhancing the fill factor and detector efficiency could further improve the measurement by an additional order of magnitude. This ground-breaking measurement will be analyzed by astronomers to extract valuable physics insights into Betelgeuse, laying the foundation for designing application-specific SPAD detectors that will reveal the mysteries of our universe.

6 Conclusions and future work

This book advances the field of SPAD-based sensors by introducing a novel vision sensor that merges SPAD detectors and event-driven data-decimation concepts in a scalable architecture. The proposed architecture's feasibility is backed by a validation ASIC that represents the first-of-its-class proof-of-concept for a new class of sensor systems. This sensor is the core of a dedicated camera module, designed and built as part of the book, with 2D and 3D imaging capabilities, event-driven motion detection, and multiple operating modes. A bottom-up characterization of the elements involved in the acquisition front-end, and the information processing chain supports verification of the proposed architectures, circuits, and methods. The bullet points below list most relevant conclusions:

- The concept of a discrete dynamic vision sensor is proposed and described, including theoretical probabilistic models supported by experimental results that define its behavior and most significant metrics. Among other features, this concept seamlessly aligns with a mostly digital implementation approach, which results in an operation that would only be limited by photon shot noise.
- Compared to frame-comparison-based techniques used in prior SPAD-based sensors for motion detection, the new concept allows a true-event-driven DVS, thus significantly reducing latency issues associated with frame-rate limitations and achieving microsecond resolution in motion detection.
- The event-driven nature of the proposed SAER sensor aligns with the inherent asynchronous behavior of SPADs, substantially reducing the vast amount of data typical of SPAD-based sensors. By transmitting events only upon detecting a sufficient photon count, the system optimizes the use of time, energy, and storage resources since pixels that do not detect meaningful information on the scene remain unread. Moreover, the time-encoding nature of events enhances the dynamic range without requiring high-bit-depth in-pixel counters.
- By granting pixels the ability to generate events upon detecting a programmable number of photons, intensity levels can be encoded in the frequency domain, thus supporting 2D image acquisition in a way compatible with asynchronous readout channel based on the AER protocol. Indeed, the SAER sensor

DOI: 10.1201/9781003427490-6

supports multiple operation modes, including ToF, that are custom-defined at the algorithm level by setting the timing of control signals.
- The sensitivity of pixels can be individually adjusted by updating the number of photons required to trigger an event. This feature allows for the implementation of online sensitivity adjustment algorithms that further optimize the bandwidth utilization of the sensor.
- SPAD photodetectors implemented within the pixels have been adapted from previous structures and shapes in 110 nm, demonstrating high PDP peaks and low DCR.
- The module camera conceived to ember the SAER sensor has been customized as a proof-of-concept for LiDAR applications. This system incorporates essential resources for biasing the sensor and SPADs, an optical emitter vital for ToF computation, and a processing module integrated into a commercial FPGA module.
- The sensor and camera system are validated under controlled lab settings and outdoors. Specifically, the sensor was tested in an astronomy application during the occultation of the star α Orionis (Betelgeuse), revealing promising outcomes in the field owing to its unique capability to record information at the single-photon level with microsecond resolution. This temporal resolution shows an improvement of orders of magnitude compared to traditional cameras in the field.

FUTURE WORK

The book contributions target low TRL (TRL metrics are used to quality the readiness level of technologies) through basic technology research and proof-of-concept chips and systems. Envisaged future challenges include:

- Engineer the SPAD device to improve NIR efficiency. PDP at NIR falls well below 3%, posing enormous challenges when measuring the ToF, as many exposures are required to gather enough statistical data.
- Improve the fill factor of the pixel by leveraging advanced technology nodes.
- Combine multiple SPADs at the pixel level to increase sensitivity and mitigate dead-time limitations, allowing for multiple counts for laser pulse and faster comparisons in the discrete DVS mode.
- Include multiple counters in the pixel to implement a multi-event TDC. This would allow scanning of multiple bins with the same exposure, improving the efficiency and time required for the system to compute ToF.
- Incorporate dynamic vision functionalities at the pixel level by including dedicated circuitry per pixel or using clusters of multiple pixels. Implementing the discrete DVS method proposed in this book in the focal plane would eliminate bottlenecks associated with collisions in the readout channel since event classification will be made before the event readout.

- Analyze the robustness of the sample-and-hold mechanism underlying the operation of analog discrete DVS for advanced technology nodes and high temperatures.
- Integrate the driver of the laser with the sensor to improve control on the emitter. Also, optimize the design of the laser to obtain the desired pulse shape.
- Design and validate a neuromorphic processor that directly records and processes spiking signals in the context of a specific application to leverage the potential of the SAER vision sensor without the need of frame reconstruction.

Of course, reaching higher technology readiness levels will also require higher resolution sensor prototypes.

References

[Afsh19] S. Afshar, T. J. Hamilton, L. Davis, A. van Schaik, and D. Delic, "Event-based Processing of Single Photon Avalanche Diode Sensors," Dec. 2019, [Online]. Available: http://arxiv.org/abs/2001.02060

[ams18] ams OSRAM, "SPL P90 Pulsed Laser Diode in Plastic Package 25 W Peak Power Datasheet," 2023, [Online]. Available: https://dammedia.osram.info/media/resource/hires/osram-dam-6189135/SPL%20PL90_EN.pdf

[Assa11] M. Assanelli, A. Ingargiola, I. Rech, A. Gulinatti, and M. Ghioni, "Photon-timing jitter dependence on injection position in single-photon avalanche diodes," *IEEE J. Quantum Electron.*, vol. 47, no. 2, pp. 151–159, Feb. 2011, doi: 10.1109/JQE.2010.2068038.

[Astr] "AstroCamp Website," [Online]. Available: https://en.astrocamp.es/intro

[Asva16] A. Asvadi, P. Girão, P. Peixoto, and U. Nunes, "3D object tracking using RGB and LiDAR data," in *2016 IEEE 19th International Conference on Intelligent Transportation Systems (ITSC)*, 2016, pp. 1255–1260. doi: 10.1109/ITSC.2016.7795718.

[Belb10] A. N. Belbachir, *Smart Cameras*. Springer, 2010. doi: 10.1007/978-1-4419-0953-4.

[Bell13] S. Bellisai, D. Bronzi, F. A. Villa, S. Tisa, A. Tosi, and F. Zappa, "Single-photon pulsed-light indirect time-of-flight 3D ranging," *Opt. Express*, vol. 21, no. 4, p. 5086, Feb. 2013, doi: 10.1364/OE.21.005086.

[Berk15] A. Berkovich, T. Datta-Chaudhuri, and P. Abshire, "A scalable 20x20 fully asynchronous SPAD-based imaging sensor with AER readout," in *2015 IEEE International Symposium on Circuits and Systems (ISCAS)*, IEEE, May 2015, pp. 1110–1113. doi: 10.1109/ISCAS.2015.7168832.

[Blai04] F. Blais, "Review of 20 years of range sensor development," *J. Electron. Imaging*, vol. 13, no. 1, p. 231, Jan. 2004, doi: 10.1117/1.1631921.

[Blit14] Blitzstein, Joseph K. and Hwang, Jessica, "Continuous random variables," in *Introduction to Probability*. CRC Press, 2014.

[Boah00] K. A. Boahen, "Point-to-point connectivity between neuromorphic chips using address events," in *IEEE Transactions on Circuits and Systems II: Analog and Digital Signal Processing*, vol. 47, no. 5, pp. 416-434, May 2000, doi: 10.1109/82.842110.

[Boah91] K. A. Boahen and A. Andreou, "A contrast sensitive silicon retina with reciprocal synapses," *Adv. Neural Inf. Process. Syst.*, vol. 4, 1991.

[Bosc01] T. Bosch, "Laser ranging: a critical review of usual techniques for distance measurement," *Opt. Eng.*, vol. 40, no. 1, p. 10, Jan. 2001, doi: 10.1117/1.1330700.

[Brag14] L. H. C. Braga *et al.*, "A fully digital 8×16 sipm array for pet applications with per-pixel tdcs and real-time energy output," *IEEE J. Solid-State Circuits*, vol. 49, no. 1, pp. 301–314, Jan. 2014, doi: 10.1109/JSSC.2013.2284351.

[Bron16] D. Bronzi, F. Villa, S. Tisa, A. Tosi, and F. Zappa, "SPAD figures of merit for photon-counting, photon-timing, and imaging applications: A review," *IEEE Sens. J.*, vol. 16, no. 1, pp. 3–12, Jan. 2016, doi: 10.1109/JSEN.2015.2483565.

[Butt20] M. Buttafava *et al.*, "SPAD-based asynchronous-readout array detectors for image-scanning microscopy," *Optica*, vol. 7, no. 7, p. 755, Jul. 2020, doi: 10.1364/optica.391726.

[Call20a] A. de la Calle-Martos *et al.*, "Sun tracker sensor for attitude control of space navigation systems," *Electron. Imaging*, vol. 32, pp. 1–7, Jan. 2020, doi: 10.2352/ISSN.2470-1173.2020.7.ISS-231.

[Cast19] M. Castello *et al.*, "A robust and versatile platform for image scanning microscopy enabling super-resolution FLIM," *Nat. Methods*, vol. 16, no. 2, pp. 175–178, Feb. 2019, doi: 10.1038/s41592-018-0291-9.

[Cecc21] F. Ceccarelli, G. Acconcia, A. Gulinatti, M. Ghioni, I. Rech, and R. Osellame, "Recent advances and future perspectives of single-photon avalanche diodes for quantum photonics applications," *Adv. Quantum Technol.*, vol. 4, no. 2, Feb. 2021, doi: 10.1002/qute.202000102.

[Chen16] Z. Cheng, D. Palubiak, X. Zheng, M. J. Deen, and H. Peng, "Impact of silicide layer on single photon avalanche diodes in a 130 nm CMOS process," *J. Phys. Appl. Phys.*, vol. 49, no. 34, p. 345105, Sep. 2016, doi: 10.1088/0022-3727/49/34/345105.

[Chen23] H.-T. Cheng, J.-S. Pan, W.-H. Lin, Y.-C. Yang, and C.-H. Wu, "Zone-addressable 20 × 20 940 nm VCSEL array with a 5-bit binary number pattern," *Opt. Lett.*, vol. 48, no. 15, p. 3937, Aug. 2023, doi: 10.1364/ol.494760.

[Choi12] J. Choi, S. Park, J. Cho, and E. Yoon, "A 1.36 μW adaptive CMOS image sensor with reconfigurable modes of operation from available energy/illumination for distributed wireless sensor network," in *2012 IEEE International Solid-State Circuits Conference*, IEEE, Feb. 2012, pp. 112–114. doi: 10.1109/ISSCC.2012.6176897.

[Choi18] W. Choi *et al.*, "A compact resistor-based CMOS temperature sensor with an inaccuracy of 0.12 °C (3σ) and a resolution FoM of 0.43 pJ· K^2 in 65-nm CMOS," *IEEE J. Solid-State Circuits*, vol. 53, no. 12, pp. 3356–3367, Dec. 2018, doi: 10.1109/JSSC.2018.2871622.

[Cili14] U. Çilingiroğlu, B. Tar, and C. Ozmen, "On-chip photovoltaic energy conversion in bulk-CMOS for indoor applications," *IEEE Trans. Circuits Syst. Regul. Pap.*, vol. 61, no. 8, pp. 2491–2504, Aug. 2014, doi: 10.1109/TCSI.2014.2304652.

[Culu03] E. Culurciello, R. Etienne-Cummings, and K. A. Boahen, "A biomorphic digital image sensor," *IEEE J. Solid-State Circuits*, vol. 38, no. 2, pp. 281–294, Feb. 2003, doi: 10.1109/JSSC.2002.807412.

[Cusi22] I. Cusini *et al.*, "Historical perspectives, state of art and research trends of SPAD arrays and their applications (part II: SPAD arrays)," *Front. Phys.*, vol. 10, Jul. 2022, doi: 10.3389/fphy.2022.906671.

[Ding23] Z. Ding *et al.*, "A multi-mode neuromorphic vision sensor with improved brightness measurement performance by pulse coding method," *IEEE Internet Things J.*, pp. 1–1, 2023, doi: 10.1109/JIOT.2023.3311019.

[Dona21] S. Donati, *Photodetectors – Devices, Circuits and Applications*, 2nd ed. Wiley-IEEE Press, 2021.

[Drag20] A. Dragan, A. Negut, A. M. Tache, and G. Brezeanu, "A reprogrammable fuse with EEcells for trimming a temperature sensor," in *2020 International Semiconductor Conference (CAS)*, IEEE, Oct. 2020, pp. 111–114. doi: 10.1109/CAS50358.2020.9268008.

[Dutt16] N. A. W. Dutton *et al.*, "A SPAD-based QVGA image sensor for single-photon counting and quanta imaging," *IEEE Trans. Electron Devices*, vol. 63, no. 1, pp. 189–196, Jan. 2016, doi: 10.1109/TED.2015.2464682.

[Elmo48] W. C. Elmore, "The transient response of damped linear networks with particular regard to wideband amplifiers," *J. Appl. Phys.*, vol. 19, no. 1, pp. 55–63, Jan. 1948, doi: 10.1063/1.1697872.

[EMVA10] EMVA, "EMVA Standard 1288 Standard for Characterization of Image Sensors and Cameras," 2010. [Online]. Available: https://api.semanticscholar.org/CorpusID:233489376

[Enz06] C. C. Enz and E. A. Vittoz, *Charge-Based MOS Transistor Modeling: The EKV Model for Low-Power and RF IC Design*, 1st ed. Wiley, 2006. Accessed: Feb. 03, 2024.

[Online]. Available: www.perlego.com/es/book/2753865/chargebased-mos-transistor-modeling-the-ekv-model-for-lowpower-and-rf-ic-design-pdf

[Enz96] C. C. Enz and G. C. Temes, "Circuit techniques for reducing the effects of op-amp imperfections: autozeroing, correlated double sampling, and chopper stabilization," *Proc. IEEE*, vol. 84, no. 11, pp. 1584–1614, 1996, doi: 10.1109/5.542410.

[Eric07] R. W. Erickson and D. Maksimovic, *Fundamentals of Power Electronics*. Springer Science & Business Media, 2007.

[Ferr19] E. Ferro, V. M. Brea, P. Lopez, and D. Cabello, "Micro-energy harvesting system including a PMU and a solar cell on the same substrate with cold startup from 2.38 nW and input power range up to 10 μW using continuous MPPT," *IEEE Trans. Power Electron.*, vol. 34, no. 6, pp. 5105–5116, Jun. 2019, doi: 10.1109/TPEL.2018.2877105.

[Fina20] T. Finateu *et al.*, "5.10 a 1280×720 back-illuminated stacked temporal contrast event-based vision sensor with 4.86μm pixels, 1.066GEPS readout, programmable event-rate controller and compressive data-formatting pipeline," in *2020 IEEE International Solid- State Circuits Conference - (ISSCC)*, IEEE, Feb. 2020, pp. 112–114. doi: 10.1109/ISSCC19947.2020.9063149.

[Font11] R. Fontaine, "Recent innovations in CMOS image sensors," in *2011 IEEE/SEMI Advanced Semiconductor Manufacturing Conference*, IEEE, May 2011, pp. 1–5. doi: 10.1109/ASMC.2011.5898219.

[Foss13] E. R. Fossum, "Modeling the performance of single-bit and multi-bit quanta image sensors," *IEEE J. Electron Devices Soc.*, vol. 1, no. 9, pp. 166–174, 2013, doi: 10.1109/JEDS.2013.2284054.

[Foss14] E. R. Fossum and D. B. Hondongwa, "A review of the pinned photodiode for CCD and CMOS image sensors," *IEEE J. Electron Devices Soc.*, vol. 2, no. 3, pp. 33–43, May 2014, doi: 10.1109/JEDS.2014.2306412.

[Foss97] E. R. Fossum, "CMOS image sensors: Electronic camera on a chip," *IEEE Trans. Electron Devices*, vol. 44, no. 10, pp. 1689–1698, 1997, doi: 10.1109/16.628824.

[Fuch10] S. Fuchs, "Multipath interference compensation in time-of-flight camera images," in *2010 20th International Conference on Pattern Recognition*, IEEE, Aug. 2010, pp. 3583–3586. doi: 10.1109/ICPR.2010.874.

[Funa15] R. Funatsu *et al.*, "6.2 133Mpixel 60fps CMOS image sensor with 32-column shared high-speed column-parallel SAR ADCs," in *2015 IEEE International Solid-State Circuits Conference - (ISSCC) Digest of Technical Papers*, IEEE, Feb. 2015, pp. 1–3. doi: 10.1109/ISSCC.2015.7062951.

[Ge11] G. Ge, C. Zhang, G. Hoogzaad, and K. A. A. Makinwa, "A single-trim CMOS bandgap reference with a 3σ Inaccuracy of ±0.15% from -40°C to 125°C," *IEEE J. Solid-State Circuits*, vol. 46, no. 11, pp. 2693–2701, Nov. 2011, doi: 10.1109/JSSC.2011.2165235.

[Gene22] T. Genevois, J. B. Horel, A. Renzaglia, and C. Laugier, "Augmented reality on LiDAR data: Going beyond vehicle-in-the-loop for automotive software validation," in *IEEE Intelligent Vehicles Symposium, Proceedings*, Institute of Electrical and Electronics Engineers Inc., 2022, pp. 971–976. doi: 10.1109/IV51971.2022.9827351.

[Genn23] G. Gennis, A. Kamperi, V. Alimisis, C. Dimas, and P. P. Sotiriadis, "An area-efficient, analog integrated image edge detector based on the Robert's cross operator," in *2023 12th International Conference on Modern Circuits and Systems Technologies (MOCAST)*, IEEE, Jun. 2023, pp. 1–4. doi: 10.1109/MOCAST57943.2023.10176648.

[Gome20] R. Gomez-Merchan, D. Palomeque-Mangut, J. A. Lenero-Bardallo, M. Delgado-Restituto, and A. Rodriguez-Vazquez, "A comparative study of stacked-diode configurations operating in the photovoltaic region," *IEEE Sens. J.*, vol. 20, no. 16, pp. 9105–9113, Aug. 2020, doi: 10.1109/JSEN.2020.2987393.

[Gome21] R. Gomez-Merchan, M. Lopez-Carmona, J. A. Lenero-Bardallo, and A. Rodriguez-Vazquez, "A high-speed low-power sun sensor with solar cells and continuous operation," in *ESSCIRC 2021 - IEEE 47th European Solid State Circuits Conference, Proceedings*, 2021. doi: 10.1109/ESSCIRC53450.2021.9567846.

[Gome22] R. Gomez-Merchan, R. D. L. Rosa-Vidal, J. A. Lenero-Bardallo, and A. Rodriguez-Vazquez, "On the implementation of in-pixel controlled diodes with sensing and energy harvesting capabilities," in *PRIME 2022 - 17th International Conference on Ph.D Research in Microelectronics and Electronics, Proceedings*, 2022. doi: 10.1109/PRIME55000.2022.9816768.

[Gome23a] R. Gomez-Merchan, J. A. Leñero-Bardallo, and Á. Rodríguez-Vázquez, "A self-powered asynchronous image sensor with TFS operation," *IEEE Sens. J.*, vol. 23, no. 7, pp. 6779–6790, Apr. 2023, doi: 10.1109/JSEN.2023.3248177.

[Gome23b] R. Gomez-Merchan, J. A. Leñero-Bardallo, and Á. Rodríguez-Vázquez, "A self-powered asynchronous image sensor with independent inpixel harvesting and sensing operations," in *IS&T International Symposium on Electronic Imaging Science and Technology*, 2023. doi: 10.2352/EI.2023.35.6.ISS-329.

[Gome23c] R. Gomez-Merchan, J. A. Lenero-Bardallo, M. Lopez-Carmona, and A. Rodriguez-Vazquez, "A low-latency, low-power CMOS sun sensor for attitude calculation using photovoltaic regime and on-chip centroid computation," *IEEE Trans. Instrum. Meas.*, vol. 72, 2023, doi: 10.1109/TIM.2023.3268478.

[Gome23d] R. Gomez-Merchan, J. A. Leñero-Bardallo, R. de L. Rosa-Vidal, and Á. Rodríguez-Vázquez, "A 64×64 SPAD-based 3D image sensor with adaptive pixel sensitivity and asynchronous readout," in *ESSCIRC 2023- IEEE 49th European Solid State Circuits Conference (ESSCIRC)*, Sep. 2023, pp. 101–104. doi: 10.1109/ESSCIRC59616.2023.10268705.

[Gome23e] R. Gomez-Merchan, R. D. L. Rosa-Vidal, J. A. Leniero-Bardallo, and A. Rodriguez-Vazquez, "Load reduction and adaptive pull-up strategies for time delay reduction in high-resolution AER sensors," in *Proceedings - IEEE International Symposium on Circuits and Systems*, 2023. doi: 10.1109/ISCAS46773.2023.10182195.

[Gott09] M. Gottardi, N. Massari, and S. A. Jawed, "A 100 μW 128 × 64 pixels contrast-based asynchronous binary vision sensor for sensor networks applications," *IEEE J. Solid-State Circuits*, vol. 44, no. 5, pp. 1582–1592, May 2009, doi: 10.1109/JSSC.2009.2017000.

[Gram22] F. Gramuglia, M. L. Wu, C. Bruschini, M. J. Lee, and E. Charbon, "A low-noise CMOS SPAD pixel with 12.1 Ps SPTR and 3 Ns dead time," *IEEE J. Sel. Top. Quantum Electron.*, vol. 28, no. 2, 2022, doi: 10.1109/JSTQE.2021.3088216.

[Guo23a] M. Guo *et al.*, "A 3-wafer-stacked hybrid 15MPixel CIS + 1 MPixel EVS with 4.6GEvent/s readout, in-pixel TDC and on-chip ISP and ESP function," in *2023 IEEE International Solid- State Circuits Conference (ISSCC)*, IEEE, Feb. 2023, pp. 90–92. doi: 10.1109/ISSCC42615.2023.10067476.

[Guo23b] S. Guo and T. Delbruck, "Low cost and latency event camera background activity denoising," *IEEE Trans. Pattern Anal. Mach. Intell.*, vol. 45, no. 1, pp. 785–795, Jan. 2023, doi: 10.1109/TPAMI.2022.3152999.

[Gyon22] I. Gyongy, N. A. W. Dutton, and R. K. Henderson, "Direct time-of-flight single-photon imaging," *IEEE Trans. Electron Devices*, vol. 69, no. 6, pp. 2794–2805, Jun. 2022, doi: 10.1109/TED.2021.3131430.

[Gyon23] I. Gyongy *et al.*, "A direct time-of-flight image sensor with in-pixel surface detection and dynamic vision," *IEEE J. Sel. Top. Quantum Electron.*, 2023, doi: 10.1109/JSTQE.2023.3238520.

[Hama20] W. Hamad, B. M. Sanayeh, M. M. Hamad, and W. H. E. Hofmann, "Impedance characteristics and chip-parasitics extraction of high-performance VCSELs," *IEEE J. Quantum Electron.*, vol. 56, no. 1, Feb. 2020, doi: 10.1109/JQE.2019.2953710.

[Hans13] M. Hansard, S. Lee, O. Choi, and R. Horaud, *Time-of-Flight Cameras*. Springer, 2013. doi: 10.1007/978-1-4471-4658-2.

[Hast05] A. Hastings, *The Art of Analog Layout*, 2nd ed. Prentice Hall, 2005.

[Hend19] R. K. Henderson *et al.*, "5.7 A 256×256 40nm/90nm CMOS 3D-stacked 120dB dynamic-range reconfigurable time-resolved SPAD imager," in *2019 IEEE International Solid- State Circuits Conference - (ISSCC)*, IEEE, Feb. 2019, pp. 106–108. doi: 10.1109/ISSCC.2019.8662355.

[Henz10] S. Henzler, *Time-to-Digital Converters*, vol. 29. Springer, 2010. doi: 10.1007/978-90-481-8628-0.

[Hutc19] S. W. Hutchings *et al.*, "A reconfigurable 3-D-stacked SPAD imager with in-pixel histogramming for flash LIDAR or high-speed time-of-flight imaging," *IEEE J. Solid-State Circuits*, vol. 54, no. 11, pp. 2947–2956, Nov. 2019, doi: 10.1109/JSSC.2019.2939083.

[icH19] ic Haus, "Datasheet of iC-HG30," [Online]. Available: www.ichaus.de/product/ic-hg30/#documents

[IEC14] *IEC 60825-1: Safety of Laser Products - Part 1: Equipment Classification and Requirements*, 4th ed. International Electrotechnical Commission (IEC), 2014.

[Jane07] Janesick, James R., "Photon transfer noise sources," in *Photon Transfer*. SPIE Press, 2007, pp. 21–34.

[Jo14] Y. R. Jo, S. K. Hong, and O. K. Kwon, "CMOS flat-panel X-ray detector with dual-gain active pixel sensors and column-parallel readout circuits," *IEEE Trans. Nucl. Sci.*, vol. 61, no. 5, pp. 2472–2479, Oct. 2014, doi: 10.1109/TNS.2014.2343459.

[Kane61] E. O. Kane, "Theory of tunneling," *J. Appl. Phys.*, vol. 32, no. 1, pp. 83–91, Jan. 1961, doi: 10.1063/1.1735965.

[Kawa07] S. Kawahito and N. Kawai, "Column parallel signal processing techniques for reducing thermal and RTS noises in CMOS image sensors," in *Proceedings in International Image Sensing Workshop*, Nov. 2007.

[Kim23] Y. Kim *et al.*, "CloudNet: A LiDAR-based face anti-spoofing model that is robust against light variation," *IEEE Access*, vol. 11, pp. 16984–16993, 2023, doi: 10.1109/ACCESS.2023.3242654.

[Koch94] C. Koch and H. Li, *Vision Chips: Implementing Vision Algorithms with Analog VLSI Circuits*. IEEE Computer Society Press, 1994.

[Koda23] K. Kodama *et al.*, "1.22μm 35.6Mpixel RGB hybrid event-based vision sensor with 4.88μm-pitch event pixels and up to 10K event frame rate by adaptive control on event sparsity," in *2023 IEEE International Solid- State Circuits Conference (ISSCC)*, IEEE, Feb. 2023, pp. 92–94. doi: 10.1109/ISSCC42615.2023.10067520.

[Kram02a] J. Kramer, "An ON/OFF transient imager with event-driven, asynchronous readout," in *2002 IEEE International Symposium on Circuits and Systems. Proceedings (Cat. No.02CH37353)*, IEEE, 2002, p. II-165-II–168. doi: 10.1109/ISCAS.2002.1010950.

[Kram02b] J. Kramer, "An integrated optical transient sensor," *IEEE Trans. Circuits Syst. II Analog Digit. Signal Process.*, vol. 49, no. 9, pp. 612–628, Sep. 2002, doi: 10.1109/TCSII.2002.807270.

[Kris15] M. Kriss, *Handbook of Digital Imaging*. Wiley, 2015.

[Kuma21] O. Kumagai *et al.*, "7.3 A 189×600 back-illuminated stacked SPAD direct time-of-flight depth sensor for automotive LiDAR systems," in *2021 IEEE International Solid- State Circuits Conference (ISSCC)*, IEEE, Feb. 2021, pp. 110–112. doi: 10.1109/ISSCC42613.2021.9365961.

[Leit13] T. Leitner *et al.*, "Measurements and simulations of low dark count rate single photon avalanche diode device in a low voltage 180-nm CMOS image sensor technology," *IEEE Trans. Electron Devices*, vol. 60, no. 6, pp. 1982–1988, 2013, doi: 10.1109/TED.2013.2259172.

[Lene16] Leñero-Bardallo, Juan Antonio and Rodríguez-Vázquez, Ángel, "ADCs for image sensors: Review and performance analysis," in *Analog Electronics for Radiation Detection*. SPIE Press, 2016, pp. 47–70.

[Lene17] J. A. Leñero-Bardallo, R. Carmona-Galan, and A. Rodriguez-Vazquez, "A wide linear dynamic range image sensor based on asynchronous self-reset and tagging of saturation events," *IEEE J. Solid-State Circuits*, vol. 52, no. 6, pp. 1605–1617, Jun. 2017, doi: 10.1109/JSSC.2017.2679058.

[Lene18] J. A. Leñero-Bardallo, R. Carmona-Galán, and A. Rodríguez-Vázquez, "Applications of event-based image sensors—Review and analysis," *Int. J. Circuit Theory Appl.*, vol. 46, no. 9, pp. 1620–1630, Sep. 2018, doi: 10.1002/cta.2546.

[Li20] Y. Li and J. Ibanez-Guzman, "Lidar for autonomous driving: The principles, challenges, and trends for automotive lidar and perception systems," *IEEE Signal Process. Mag.*, vol. 37, no. 4, pp. 50–61, Jul. 2020, doi: 10.1109/MSP.2020.2973615.

[Lich06] P. Lichtsteiner, C. Posch, and T. Delbruck, "A 128 X 128 120db 30mw asynchronous vision sensor that responds to relative intensity change," in *2006 IEEE International Solid State Circuits Conference - Digest of Technical Papers*, IEEE, 2006, pp. 2060–2069. doi: 10.1109/ISSCC.2006.1696265.

[Lich08] P. Lichtsteiner, C. Posch, and T. Delbruck, "A 128 × 128 120 dB 15 μs latency asynchronous temporal contrast vision sensor," *IEEE J. Solid-State Circuits*, vol. 43, no. 2, pp. 566–576, Feb. 2008, doi: 10.1109/JSSC.2007.914337.

[Liu98] S. Liu and R. J. Baker, "Process and temperature performance of a CMOS beta-multiplier voltage reference," *1998 Midwest Symposium on Circuits and Systems (Cat. No. 98CB36268)*, IEEE, 1998, pp. 33–36, doi: 10.1109/MWSCAS.1998.759429.

[Lope18] J. M. Lopez-Martinez, I. Vornicu, R. Carmona-Galan, and A. Rodriguez-Vazquez, "An experimentally-validated Verilog-A SPAD model extracted from TCAD simulation," in *2018 25th IEEE International Conference on Electronics, Circuits and Systems (ICECS)*, IEEE, Dec. 2018, pp. 137–140. doi: 10.1109/ICECS.2018.8617962.

[Lopi11] A. Lopich and P. Dudek, "A SIMD cellular processor array vision chip with asynchronous processing capabilities," *IEEE Trans. Circuits Syst. Regul. Pap.*, vol. 58, no. 10, pp. 2420–2431, Oct. 2011, doi: 10.1109/TCSI.2011.2131370.

[Lott11] C. Lotto, P. Seitz, and T. Baechler, "A sub-electron readout noise CMOS image sensor with pixel-level open-loop voltage amplification," in *2011 IEEE International Solid-State Circuits Conference*, IEEE, Feb. 2011, pp. 402–404. doi: 10.1109/ISSCC.2011.5746370.

[Main13] A. K. Maini, *Lasers and Optoelectronics: Fundamentals, Devices and Applications*. John Wiley & Sons, 2013.

[McKe66] J. P. McKelvey, *Solid State and Semiconductor Physics*. Harper & Row, 1966.

[Mead88] C. A. Mead and M. A. Mahowald, "A silicon model of early visual processing," *Neural Netw.*, vol. 1, no. 1, pp. 91–97, Jan. 1988, doi: 10.1016/0893-6080(88)90024-X.

[Meyn09] G. Meynants, G. Lepage, J. Bogaerts, G. Vanhorebeek, and X. Wang, "Limitations to the frame rate of high speed image sensors," in *2009 International Image Sensor Workshop (IISW)*, 2009, pp. 153–156.

[Miro07] L. Miró-Amarante *et al.*, "LVDS serial AER link performance," in *2007 IEEE International Symposium on Circuits and Systems (ISCAS)*, 2007, pp. 1537–1540. doi: 10.1109/ISCAS.2007.378704.

[Mont21] M. Montargès *et al.*, "A dusty veil shading Betelgeuse during its great dimming," *Nature*, vol. 594, no. 7863, pp. 365–368, Jun. 2021, doi: 10.1038/s41586-021-03546-8.

[More19] M. Moreno-Garcia, L. Pancheri, M. Perenzoni, R. del Rio, O. G. Vinuesa, and A. Rodriguez-Vazquez, "characterization-based modeling of retriggering and afterpulsing for passively quenched CMOS SPADs," *IEEE Sens. J.*, vol. 19, no. 14, pp. 5700–5709, Jul. 2019, doi: 10.1109/JSEN.2019.2903937.

[Mori20a] K. Morimoto *et al.*, "Megapixel time-gated SPAD image sensor for 2D and 3D imaging applications," *Optica*, vol. 7, no. 4, p. 346, Apr. 2020, doi: 10.1364/optica.386574.

[Mori20b] K. Morimoto and E. Charbon, "High fill-factor miniaturized SPAD arrays with a guard-ring-sharing technique," *Opt. Express*, vol. 28, no. 9, p. 13068, Apr. 2020, doi: 10.1364/OE.389216.

[Naka06] J. Nakamura, *Image Sensors and Signal Processing for Digital Still Cameras*. Taylor & Francis, 2006.

[Ni11] Y. Ni, Y. Zhu and B. Arion, "A 768x576 logarithmic image sensor with photodiode in solar cell mode," in *2011 International Image Sensor Workshop (IISW)*, IEEE, 2011.

[Nicl07] C. Niclass, M. Gersbach, R. Henderson, L. Grant, and E. Charbon, "A single photon avalanche diode implemented in 130-nm CMOS technology," *IEEE J. Sel. Top. Quantum Electron.*, vol. 13, no. 4, pp. 863–869, Jul. 2007, doi: 10.1109/JSTQE.2007.903854.

[Nicl10] C. Niclass, M. Soga, and S. Kato, "A 0.18µm CMOS single-photon sensor for coaxial laser rangefinders," in *2010 IEEE Asian Solid-State Circuits Conference*, IEEE, Nov. 2010, pp. 1–4. doi: 10.1109/ASSCC.2010.5716568.

[Nicl13] C. Niclass, M. Soga, H. Matsubara, S. Kato, and M. Kagami, "A 100-m range 10-frame/s 340x96-pixel time-of-flight depth sensor in 0.18-µm CMOS," *IEEE J. Solid-State Circuits*, vol. 48, no. 2, pp. 559–572, Feb. 2013, doi: 10.1109/JSSC.2012.2227607.

[Niwa23] A. Niwa *et al.*, "A 2.97µm-pitch event-based vision sensor with shared pixel front-end circuitry and low-noise intensity readout mode," in *Digest of Technical Papers - IEEE International Solid-State Circuits Conference*, Institute of Electrical and Electronics Engineers Inc., 2023, pp. 94–96. doi: 10.1109/ISSCC42615.2023.10067566.

[Nobl68] P. J. W. Noble, "Self-scanned silicon image detector arrays," *IEEE Trans. Electron Devices*, vol. 15, no. 4, pp. 202–209, Apr. 1968, doi: 10.1109/T-ED.1968.16167.

[Ogi21] J. Ogi *et al.*, "7.5 A 250fps 124dB dynamic-range SPAD image sensor stacked with pixel-parallel photon counter employing sub-frame extrapolating architecture for motion artifact suppression," in *2021 IEEE International Solid- State Circuits Conference (ISSCC)*, IEEE, Feb. 2021, pp. 113–115. doi: 10.1109/ISSCC42613.2021.9365977.

[Ohta07] J. Ohta, *Smart CMOS Image Sensors and Applications*, 1st ed. CRC Press, 2007.

[Okad21] C. Okada *et al.*, "7.6 A high-speed back-illuminated stacked CMOS image sensor with column-parallel kT/C-cancelling S&H and delta-sigma ADC," in *2021 IEEE International Solid- State Circuits Conference (ISSCC)*, IEEE, Feb. 2021, pp. 116–118. doi: 10.1109/ISSCC42613.2021.9366024.

[Orti23] J. L. Ortiz *et al.*, "The stellar occultation by (319) Leona on 2023 September 13 in preparation for the occultation of Betelgeuse," *Mon. Not. R. Astron. Soc. Lett.*, vol. 528, no. 1, pp. L139–L145, Nov. 2023, doi: 10.1093/mnrasl/slad179.

[Ouh20] H. Ouh, B. Shen, and M. L. Johnston, "Combined in-pixel linear and single-photon avalanche diode operation with integrated biasing for wide-dynamic-range optical sensing," *IEEE J. Solid-State Circuits*, vol. 55, no. 2, pp. 392–403, Feb. 2020, doi: 10.1109/JSSC.2019.2944856.

[Padm21] P. Padmanabhan *et al.*, "7.4 A 256×128 3D-stacked (45nm) SPAD FLASH LiDAR with 7-level coincidence detection and progressive gating for 100m range and 10klux background light," in *2021 IEEE International Solid- State Circuits Conference (ISSCC)*, IEEE, Feb. 2021, pp. 111–113. doi: 10.1109/ISSCC42613.2021.9366010.

[Palu14] D. P. Palubiak and M. J. Deen, "CMOS SPADs: Design issues and research challenges for detectors, circuits, and arrays," *IEEE J. Sel. Top. Quantum Electron.*, vol. 20, no. 6, pp. 409–426, Nov. 2014, doi: 10.1109/JSTQE.2014.2344034.

[Pan18] S. Pan and K. A. A. Makinwa, "A 0.25mm^2 resistor-based temperature sensor with an inaccuracy of 0.12°C (3σ) from -55°C to 125°C and a resolution FOM of 32fJ·K^2," in *2018 IEEE International Solid - State Circuits Conference - (ISSCC)*, IEEE, Feb. 2018, pp. 320–322. doi: 10.1109/ISSCC.2018.8310313.

[Pan19] G. Pan *et al.*, "Ultra-compact electrically controlled beam steering chip based on coherently coupled VCSEL array directly integrated with optical phased array," *Opt. Express*, vol. 27, no. 10, p. 13910, May 2019, doi: 10.1364/oe.27.013910.

[Panc13] L. Pancheri, E. Panina, G.-F. D. Betta, L. Gasparini, and D. Stoppa, "Compact analog counting SPAD pixel with 1.9% PRNU and 530ps time gating," in *2013 Proceedings of the ESSCIRC (ESSCIRC)*, IEEE, Sep. 2013, pp. 295–298. doi: 10.1109/ESSCIRC.2013.6649131.

[Park18] S.-Y. Park, K. Lee, H. Song, and E. Yoon, "Simultaneous imaging and energy harvesting in CMOS image sensor pixels," *IEEE Electron Device Lett.*, vol. 39, no. 4, pp. 532–535, Apr. 2018, doi: 10.1109/LED.2018.2811342.

[Park21] B. Park *et al.*, "A 64 × 64 SPAD-based indirect time-of-flight image sensor with 2-tap analog pulse counters," *IEEE J. Solid-State Circuits*, vol. 56, no. 10, pp. 2956–2967, Oct. 2021, doi: 10.1109/JSSC.2021.3094524.

[Park23] M.-J. Park and H.-J. Kim, "A real-time edge-detection CMOS Image sensor for machine vision applications," *IEEE Sens. J.*, vol. 23, no. 9, pp. 9254–9261, May 2023, doi: 10.1109/JSEN.2023.3263461.

[Peli90] E. Peli, "Contrast in complex images," *J. Opt. Soc. Am. A*, vol. 7, no. 10, p. 2032, Oct. 1990, doi: 10.1364/JOSAA.7.002032.

[Peng23] Z. Peng, Z. Xiong, Y. Zhao, and L. Zhang, "3-D objects detection and tracking using solid-state LiDAR and RGB camera," *IEEE Sens. J.*, vol. 23, no. 13, pp. 14795–14808, Jul. 2023, doi: 10.1109/JSEN.2023.3279500.

[Pere16] M. Perenzoni, N. Massari, D. Perenzoni, L. Gasparini, and D. Stoppa, "A 160x120 pixel analog-counting single-photon imager with time-gating and self-referenced column-parallel A/D conversion for fluorescence lifetime imaging," *IEEE J. Solid-State Circuits*, vol. 51, no. 1, pp. 155–167, Jan. 2016, doi: 10.1109/JSSC.2015.2482497.

[Pere18] M. Perenzoni, "Single-photon avalanche diode-based detection and imaging: bringing the photodiode out of its comfort zone," *IEEE Solid-State Circuits Mag.*, vol. 10, no. 3, pp. 26–34, Jun. 2018, doi: 10.1109/MSSC.2018.2844602.

[Piem19] C. Piemonte and A. Gola, "Overview on the main parameters and technology of modern Silicon Photomultipliers," *Nucl. Instrum. Methods Phys. Res. Sect. Accel. Spectrometers Detect. Assoc. Equip.*, vol. 926, pp. 2–15, May 2019, doi: 10.1016/j.nima.2018.11.119.

[Purv17] Purves, Dale *et al.*, "Vision: The eye," in *Neuroscience*, 6th ed.. OUP, 2017.

[Raza17] B. Razavi, *Design Of Analog CMOS Integrated Circuit*, 2nd ed. Mc Graw Hill, 2017.

[Rocc20] F. M. D. Rocca *et al.*, "A 128 × 128 SPAD motion-triggered time-of-flight image sensor with in-pixel histogram and column-parallel vision processor," *IEEE J. Solid-State Circuits*, vol. 55, no. 7, pp. 1762–1775, Jul. 2020, doi: 10.1109/JSSC.2020.2993722.

[Rodr18] A. Rodríguez-Vázquez, J. Fernández-Berni, J. A. Leñero-Bardallo, I. Vornicu, and R. Carmona-Galán, "CMOS vision sensors: Embedding computer vision at imaging front-ends," *IEEE Circuits Syst. Mag.*, vol. 18, no. 2, pp. 90–107, Apr. 2018, doi: 10.1109/MCAS.2018.2821772.

[Rodr23] Á. Rodriguez-Vázquez, "ESSCIRC 2023 keynote speakers: Chip architectures for efficient analog-to-information image analysis using out-the-box processing concepts," in *ESSCIRC 2023- IEEE 49th European Solid State Circuits Conference (ESSCIRC)*, IEEE, Sep. 2023, pp. i–iv. doi: 10.1109/ESSCIRC59616.2023.10268700.

[Rosk01] T. Roska and Á. Rodríguez-Vázquez, *Towards the Visual Microprocessor*. John Wiley & Sons, 2001.

[Rosk06] B. Roska, A. Molnar, and F. S. Werblin, "Parallel processing in retinal ganglion cells: How integration of space-time patterns of excitation and inhibition form the spiking output," *J. Neurophysiol.*, vol. 95, no. 6, pp. 3810–3822, Jun. 2006, doi: 10.1152/jn.00113.2006.

[Rosk93] T. Roska and L. O. Chua, "The CNN universal machine: an analogic array computer," *IEEE Trans. Circuits Syst. II Analog Digit. Signal Process.*, vol. 40, no. 3, pp. 163–173, Mar. 1993, doi: 10.1109/82.222815.

[Saka18] M. Sakakibara *et al.*, "A back-illuminated global-shutter CMOS image sensor with pixel-parallel 14b subthreshold ADC," in *2018 IEEE International Solid - State Circuits Conference - (ISSCC)*, IEEE, Feb. 2018, pp. 80–82. doi: 10.1109/ISSCC.2018.8310193.

[Saku90] T. Sakurai and A. R. Newton, "Alpha-power law MOSFET model and its applications to CMOS inverter delay and other formulas," *IEEE J. Solid-State Circuits*, vol. 25, no. 2, pp. 584–594, Apr. 1990, doi: 10.1109/4.52187.

[Sato20] M. Sato *et al.*, "5.8 A 0.50e- rms Noise 1.45μm-pitch CMOS image sensor with reference-shared in-pixel differential amplifier at 8.3Mpixel 35fps," in *2020 IEEE International Solid- State Circuits Conference - (ISSCC)*, IEEE, Feb. 2020, pp. 108–110. doi: 10.1109/ISSCC19947.2020.9063017.

[Sego17] J. A. Segovia, F. Medeiro, A. González, A. Villegas, and Á. Rodríguez-Vázquez, "A 5-megapixel 100-frames-per-second 0.5erms low noise CMOS image sensor with column-parallel two-stage oversampled analog-to-digital converter," in *2017 International Image Sensor Workshop*, 2017.

[Seit13] P. Seitz and A. Theuwissen, *Single-Photon Imaging*. Springer, 2013.

[Seve23] F. Severini *et al.*, "spatially resolved event-driven 24 × 24 pixels SPAD imager with 100% duty cycle for low optical power quantum entanglement detection," *IEEE J. Solid-State Circuits*, vol. 58, no. 8, pp. 2278–2287, Aug. 2023, doi: 10.1109/JSSC.2023.3249122.

[Sham18] M. Shamim, A. Shawkat, and N. McFarlane, "A CMOS perimeter gated SPAD based mini-digital silicon photomultiplier," in *2018 IEEE 61st International Midwest Symposium on Circuits and Systems (MWSCAS)*, IEEE, Aug. 2018, pp. 302–305. doi: 10.1109/MWSCAS.2018.8624054.

[Shaw23] M. S. A. Shawkat, M. M. Adnan, R. D. Febbo, J. J. Murray, and G. S. Rose, "A single chip SPAD based vision sensing system with integrated memristive spiking neuromorphic processing," *IEEE Access*, vol. 11, pp. 19441–19457, 2023, doi: 10.1109/ACCESS.2023.3244793.

[Shir20] Y. Shirakawa, K. Yasutomi, K. Kagawa, S. Aoyama, and S. Kawahito, "An 8-tap CMOS lock-in pixel image sensor for short-pulse time-of-flight measurements," *Sensors*, vol. 20, no. 4, p. 1040, Feb. 2020, doi: 10.3390/s20041040.

[Shoc52] W. Shockley and W. T. Read, "Statistics of the recombinations of holes and electrons," *Phys. Rev.*, vol. 87, no. 5, pp. 835–842, Sep. 1952, doi: 10.1103/PhysRev.87.835.

[Sido02] D. N. Sidorov and A. C. Kokaram, "Suppression of moire patterns via spectral analysis," C.-C. J. Kuo, Ed., Jan. 2002, p. 895. doi: 10.1117/12.453134.

[Sing13] J. Singh, S. P. Mohanty, and D. K. Pradhan, *Robust SRAM Designs and Analysis*. Springer, 2013. doi: 10.1007/978-1-4614-0818-5.

[Soel16] C. Soell, L. Shi, J. Roeber, M. Reichenbach, R. Weigel, and A. Hagelauer, "Low-power analog smart camera sensor for edge detection," in *2016 IEEE International*

Conference on Image Processing (ICIP), IEEE, Sep. 2016, pp. 4408–4412. doi: 10.1109/ICIP.2016.7533193.

[Son17] B. Son *et al.*, "4.1 A 640×480 dynamic vision sensor with a 9μm pixel and 300Meps address-event representation," in *2017 IEEE International Solid-State Circuits Conference (ISSCC)*, IEEE, Feb. 2017, pp. 66–67. doi: 10.1109/ISSCC.2017.7870263.

[Spin97] A. Spinelli and A. L. Lacaita, "Physics and numerical simulation of single photon avalanche diodes," *IEEE Trans. Electron Devices*, vol. 44, no. 11, pp. 1931–1943, 1997, doi: 10.1109/16.641363.

[Suar17] M. Suarez, V. M. Brea, J. Fernandez-Berni, R. Carmona-Galan, D. Cabello, and A. Rodriguez-Vazquez, "Low-power CMOS vision sensor for Gaussian pyramid extraction," *IEEE J. Solid-State Circuits*, vol. 52, no. 2, pp. 483–495, Feb. 2017, doi: 10.1109/JSSC.2016.2610580.

[Suh20] Y. Suh *et al.*, "A 1280×960 dynamic vision sensor with a 4.95-μm pixel pitch and motion artifact minimization," in *2020 IEEE International Symposium on Circuits and Systems (ISCAS)*, IEEE, Oct. 2020, pp. 1–5. doi: 10.1109/ISCAS45731.2020.9180436.

[Sze06] S. M. Sze and K. K. Ng, *Physics of Semiconductor Devices*. Wiley, 2006. doi: 10.1002/0470068329.

[Tane22] F. Taneski, T. A. Abbas, and R. K. Henderson, "Laser power efficiency of partial histogram direct time-of-flight LiDAR sensors," *J. Light. Technol.*, vol. 40, no. 17, pp. 5884–5893, Sep. 2022, doi: 10.1109/JLT.2022.3187293.

[Tang18] F. Tang *et al.*, "An Area-efficient column-parallel digital decimation filter with pre-BWI topology for CMOS image sensor," *IEEE Trans. Circuits Syst. Regul. Pap.*, vol. 65, no. 8, pp. 2524–2533, Aug. 2018, doi: 10.1109/TCSI.2018.2795086.

[Tele] Teledyne Photometrics, "Rolling vs Global Shutter," [Online]. Available: www.photometrics.com/learn/white-papers/rolling-vs-global-shutter

[Tele18] Teledyne e2v, "Datasheet of eye-RIS vSoC," [Online]. Available: https://imaging.teledyne-e2v.com/content/uploads/2019/01/33613-AnaFocus_Eye-RIS-VSoC_v3_AW_WEB.pdf

[Tera12] N. Teranishi, "Required conditions for photon-counting image sensors," *IEEE Trans. Electron Devices*, vol. 59, no. 8, pp. 2199–2205, 2012, doi: 10.1109/TED.2012.2200487.

[Tetr15] M. A. Tétrault *et al.*, "Real-time discrete SPAD array readout architecture for time of flight PET," *IEEE Trans. Nucl. Sci.*, vol. 62, no. 3, pp. 1077–1082, Jun. 2015, doi: 10.1109/TNS.2015.2409783.

[Torr09] A. Torralba, "How many pixels make an image?" *Vis. Neurosci.*, vol. 26, no. 1, pp. 123–131, Jan. 2009, doi: 10.1017/S0952523808080930.

[Tots16] H. Totsuka *et al.*, "6.4 An APS-H-size 250Mpixel CMOS image sensor using column single-slope ADCs with dual-gain amplifiers," in *2016 IEEE International Solid-State Circuits Conference (ISSCC)*, IEEE, Jan. 2016, pp. 116–117. doi: 10.1109/ISSCC.2016.7417934.

[Ulku19] A. C. Ulku *et al.*, "A 512 × 512 SPAD image sensor with integrated gating for widefield FLIM," *IEEE J. Sel. Top. Quantum Electron.*, vol. 25, no. 1, Jan. 2019, doi: 10.1109/JSTQE.2018.2867439.

[Varg15] S. Vargas-Sierra, G. Linán-Cembrano, and Á. Rodríguez-Vázquez, "A 151 dB high dynamic range CMOS image sensor chip architecture with tone mapping compression embedded in-pixel," *IEEE Sens. J.*, vol. 15, no. 1, pp. 180–195, Jan. 2015, doi: 10.1109/JSEN.2014.2340875.

[Veer15] C. Veerappan and E. Charbon, "CMOS SPAD based on photo-carrier diffusion achieving PDP >40% from 440 to 580 nm at 4 v excess bias," *IEEE Photonics Technol. Lett.*, vol. 27, no. 23, pp. 2445–2448, Dec. 2015, doi: 10.1109/LPT.2015.2468067.

[Vixa19] Vixar, "VCSEL pulse driver designs for ToF applications, application note," [Online]. Available: https://vixarinc.com/wp-content/uploads/2020/11/Vixar_DriverDesign_AppNote.pdf

[Vorn16] I. Vornicu, R. Carmona-Galán, B. Pérez-Verdú, and Á. Rodríguez-Vázquez, "Compact CMOS active quenching/recharge circuit for SPAD arrays," *Int. J. Circuit Theory Appl.*, vol. 44, no. 4, pp. 917–928, Apr. 2016, doi: 10.1002/cta.2113.

[Vorn17] I. Vornicu, R. Carmona-Galan, and A. Rodriguez-Vazquez, "Arrayable voltage-controlled ring-oscillator for direct time-of-flight image sensors," *IEEE Trans. Circuits Syst. Regul. Pap.*, vol. 64, no. 11, pp. 2821–2834, Nov. 2017, doi: 10.1109/TCSI.2017.2706324.

[Vorn19] I. Vornicu, A. Darie, R. Carmona-Galan, and A. Rodriguez-Vazquez, "Compact Real-time inter-frame histogram builder for 15-bits high-speed ToF-imagers based on single-photon detection," *IEEE Sens. J.*, vol. 19, no. 6, pp. 2181–2190, Mar. 2019, doi: 10.1109/JSEN.2018.2885960.

[Vorn20] I. Vornicu, F. N. Bandi, R. Carmona-Galán, and Á. Rodríguez-Vázquez, "Compact macro-cell with or pulse combining for low power digital-SiPM," *IEEE Sens. J.*, vol. 20, no. 21, pp. 12817–12826, Nov. 2020, doi: 10.1109/JSEN.2020.3002609.

[Vorn21] I. Vornicu, J. M. Lopez-Martinez, F. N. Bandi, R. C. Galan, and A. Rodriguez-Vazquez, "Design of high-efficiency SPADs for LiDAR applications in 110nm CIS technology," *IEEE Sens. J.*, vol. 21, no. 4, pp. 4776–4785, Feb. 2021, doi: 10.1109/JSEN.2020.3032106.

[Wan20] X. Wan, L. Zhang, J. Liu, J. Liu, and D. Fu, "A simple and efficient fuse-trimming circuit for analog design," in *2020 IEEE 15th International Conference on Solid-State & Integrated Circuit Technology (ICSICT)*, IEEE, Nov. 2020, pp. 1–3. doi: 10.1109/ICSICT49897.2020.9278016.

[Wang05] Wang, Zhou and Bovik, Alan C., "Foveated image and video coding," in *Digital Video Image Quality and Perceptual Coding*. CRC Press, 2005.

[Wang10] X. Wang *et al.*, "A 2.2M CMOS image sensor for high-speed machine vision applications," E. Bodegom and V. Nguyen, Eds., Feb. 2010, p. 75360M. doi: 10.1117/12.838880.

[Wang18] R. Wang *et al.*, "A Sub-1ppm/°C current-mode CMOS bandgap reference with piecewise curvature compensation," *IEEE Trans. Circuits Syst. Regul. Pap.*, vol. 65, no. 3, pp. 904–913, Mar. 2018, doi: 10.1109/TCSI.2017.2771801.

[Weit06] C. Weitkamp, *Lidar: Range-Resolved Optical Remote Sensing of the Atmosphere*, vol. 102. Springer Science & Business, 2006.

[Wuu22] S.-G. Wuu, H.-L. Chen, H.-C. Chien, P. Enquist, R. M. Guidash, and J. McCarten, "A review of 3-dimensional wafer level stacked backside illuminated CMOS image sensor process technologies," *IEEE Trans. Electron Devices*, vol. 69, no. 6, pp. 2766–2778, Jun. 2022, doi: 10.1109/TED.2022.3152977.

[Xie23] T. Xie and C. Seals, "Design of mobile augmented reality assistant application via deep learning and LIDAR for visually impaired," in *Digest of Technical Papers – IEEE International Conference on Consumer Electronics*, Institute of Electrical and Electronics Engineers Inc., 2023. doi: 10.1109/ICCE56470.2023.10043516.

[Xu17] Y. Xu, P. Xiang, and X. Xie, "Comprehensive understanding of dark count mechanisms of single-photon avalanche diodes fabricated in deep sub-micron CMOS technologies," *Solid-State Electron.*, vol. 129, pp. 168–174, Mar. 2017, doi: 10.1016/j.sse.2016.11.009.

[Yadi04] O. Yadid-Pecht and R. Etienne-Cummings, *CMOS Imagers: From Phototransduction to Image Processing*. Springer, 2004. doi: 10.1007/b117398.

[Yang22] T. Yang *et al.*, "3D ToF LiDAR in Mobile Robotics: A Review," Feb. 2022, [Online]. Available: http://arxiv.org/abs/2202.11025

[Yole23a] Yole Intelligence, "Status of the CMOS Image Sensor Industry 2023," [Online]. Available: www.yolegroup.com/product/report/status-of-the-cmos-image-sensor-industry-2023/

[Yole23b] Yole Intelligence, "3D Imaging & Sensing 2023," [Online]. Available: www.yolegroup.com/product/report/3d-imaging--sensing-2023/

[Zara11] À. Zarándy, *Focal-Plane Sensor-Processor Chips*. Yole Intelligence, 2011.

[Zhan19a] C. Zhang, S. Lindner, I. M. Antolovic, J. M. Pavia, M. Wolf, and E. Charbon, "A 30-frames/s, 252 × 144 SPAD Flash LiDAR with 1728 dual-clock 48.8-ps TDCs, and pixel-wise integrated histogramming," *IEEE J. Solid-State Circuits*, vol. 54, no. 4, pp. 1137–1151, Apr. 2019, doi: 10.1109/JSSC.2018.2883720.

[Zhan19b] X. Zhang, W. Fan, J. Xi, and L. He, "A 14-bit 150KS/s SAR ADC with PGA for CMOS image sensor," in *2019 IEEE International Symposium on Signal Processing and Information Technology (ISSPIT)*, IEEE, Dec. 2019, pp. 1–5. doi: 10.1109/ISSPIT47144.2019.9001750.

[Zhao06] L. Zhaoping, "Theoretical understanding of the early visual processes by data compression and data selection," *Netw. Comput. Neural Syst.*, vol. 17, no. 4, pp. 301–334, Jan. 2006, doi: 10.1080/09548980600931995.

Index

S

T

V

For Product Safety Concerns and Information please contact our EU representative GPSR@taylorandfrancis.com Taylor & Francis Verlag GmbH, Kaufingerstraße 24, 80331 München, Germany